FROM THEORY TO CLASSROOM TO FUN

How One Science Teacher Found Nirvana

by

Eric J Burtson

Table of Contents

Chapter 22: The End/The Beginning

Cover design by C. Michael Pedersen

ISBN: 978-0-9909287-4-4

To Daryl Ferguson
and
Bob and Marilyn Stout

FOREWORD BY ALLISON AND JOSE

It's always awkward to ask a colleague to write a foreword, because other teachers rarely see you in action. Much of what they know about you is what they hear. And though I could have gone that route, I decided instead to give you the perspective of people who have actually been in my classroom, my students.

Since I returned to teaching three years ago, I have received a few letters from students, usually on the last day of class. No one asked them to write these. I am often surprised and delighted at the sincerity of what they write.

I chose a couple of the more descriptive notes. It is through these that I have Allison and José writing the foreword. I'm sure they never imagined they would be used this way! To avoid embarrassing them, I have changed their names.

I chose these two letters because they show how good teaching changes lives (or at least, career paths). Moreover, these notes provide evidence that *fun* and *theory* combine powerfully to generate effective teaching and learning.

From Allison, a freshman in physics:

> Dear Mr. Burtson,
>
> I just wanted to thank you for all you've taught us. I can say, without a doubt that you are the best teacher I've ever had, and this includes a lady who's known me since I was 2. Thank you for inspiring me to pursue science, whichever it may be, and thank you for caring. There are some teachers who treat their students like cans in a factory, they stuff us with knowledge and send us out into the world.

Thank you for not being that sort of teacher, for caring about our hobbies, quirks, and characters. Thank you for all of this and have a great summer!

From José, a junior in physics:

I wanted to write this because I felt like your teaching techniques are the best out of every teacher I've ever had. You have fun while teaching which makes it fun for us. It's so easy to pay attention & understand the material. Your class isn't boring like other classes that make you fall asleep. I had so much fun in your class this year, so much that I'm going to miss it. (I've never been able to say that before.) Not only did I have fun but I learned so much, you've made it so I WANT to learn more and more about physics. I know you probably hear this a lot but you deserve to hear it even more. You are the best teacher out there, and I hope your future students enjoy and have as much fun in this class as I did. Thank you for making this class as fun and enjoyable as possible. Keep doing what you're doing! Have a great summer Mr. Burtson.

Sincerely,

The student who now loves Physics

I read José's note before the end of class, so I thanked him for what he said. I told him that I always appreciate hearing that I make a difference.

This exchange must have given him the freedom to expand on what he said in the letter. He continued. He told me he always hoped he could be a journeyman electrician. And now, after doing so well in physics, he felt he was smart enough to do it. I acknowledged that, but told him emphatically that he could go beyond that and be an electrical engineer. He said he would consider that.

INTRODUCTION

They say science teachers are in high demand. In my district they can't hire enough physics teachers. That might give you as a new teacher some sense of job security, but is it going to make you love your job? Is it going to make students excited to come to your classroom? Is it going to make you feel competent and powerful?

When I left engineering to teach, it felt like I made a horrible mistake. It was the kind of mistake that is so bad that all the blood rushes out of your face and leaves you nauseous. But I stuck with it. And it turned out to be the best professional decision I could ever make.

My intention for this book is to accelerate you quickly past the difficult times, and land you gently in pleasant pastures, then to teach you how to build your own rocket from sticks, stones, and clay. You can see this goal in the title of chapter 1: "Be a Top 1% Science Teacher from Day One."

Theory to Classroom to Fun is an excellent companion to and extension of what teacher credentialing programs show future teachers. It also reveals some important truths the programs might avoid getting into.

So if you are thinking of teaching, or if you have started your teacher education—even if you are one, two, or three years in—this book will give you the perspective you have been hungry for.

Chapters include

- Be a Top 1% Science Teacher from Day One (Chapter 1)
- Fun (3)
- Motivation and Engagement (4)
- Educational Theory (5)

- Your Day (6)
- Control the Class/ Lead the Students/ Know Your Kids (7)
- Planning the Year and the Lesson (8)
- Administration (9)
- Persona (10)
- Parents (17)
- Halloween (19)
- Burn Out (21)

The chapters are easy to read and filled with true tales from the classroom. They hit on important ideas.

For example, you've probably been exposed to lots of *educational theory*. Has it stood the test of time? I report on those methods that have worked well for me, in addition to highly regarded and widely implemented ideas on how best to teach. That chapter also touches on the Next Generation Science Standards.

Other chapters will help you avoid pitfalls. The *Grading* chapter will give you practical methods to save you hours of time each week. You will find other ways to be more efficient, more relevant, and maximally effective. You will learn how to make an environment that is fun for you and fun for your students.

My perspective comes from various careers within education, and from twenty-one years of informative trial and error in the classroom. I have seen things from the viewpoints of. . .

- an ex-engineer
- a young, struggling first year teacher at an urban middle school
- a new teacher whose first evaluation was not good
- a person who set his sights on a dream teaching job and got it
- a person who taught teachers in summer institutes

- a teacher who was repeatedly recruited
- a person who burned out
- a teacher who became an administrator and evaluated other teachers
- a person who left education entirely
- a person who came back to teaching with renewed enthusiasm, perspective, and effectiveness
- a teacher who has grown to appreciate the students more than the subject

What this list doesn't capture are the extremes of my experience. I went from being bad to being good. But even after I earned a strong reputation, and despite the fact that my students earned exceptional test scores, I still had trouble with a particular principal. I learned a lot from all the good and all the bad. I pass that on to you.

Still, I eventually left teaching to try other things. When I came back I was convinced, finally, that I was in the right place. I felt powerful, respected, and happy. My professional life was fun. I want to share that with you. I want you to know how it can be right for you too, or wrong—no doubt, just clarity.

Here are some of my achievements: In 2003 I was a PAEMST finalist for California (Presidential Science and Math Teaching Award). I was one of the first science teachers in my region to get National Certification. The San Diego Science Educators Association awarded me high school teacher of the year for work I did co-founding the SciTech Academy.

Finally, they say camp counselors often make excellent teachers. The problem is people who understand science often do not make good camp counselors. You might know yourself to be strong on content and weak on connection. And yet something calls you to teach. Do you belong in the classroom? Will your students learn great things from you?

I would have made a horrible camp counselor, yet I am teaching what I want. And I am teaching it how I want to. It is my dream job. My kids love learning, and I love teaching them. It is stimulating, and deeply satisfying. We talk about Hawking's proclamation that we need to colonize the universe. We share sadness over the death of Leonard Nimoy.

Yes, this is my dream job. I'm living it right now. Let me show you how.

1. Be a Top 1% Science Teacher from Day One

KNOW THYSELF

You are weird. At least the kids think so. But that's all right. The sooner you embrace this, the sooner you can work around it and use it to your advantage. Here is a story to illustrate how weird you are.

A few years ago I had a great job at a nice school. I taught physics and calculus and had a couple of periods of extra prep time to design professional development for the staff. But I left it to participate in the small school movement. I went to an urban school where my friend Steve was principal.

Things were different at this school. One day a student stayed after class to tell me something. Jack had piercings and tattoos and a certain swagger to his strut. He said, "Mr. Burtson, you always give us homework, and no one ever does it, not even me, and I'm your best student."

It's true, Jack was clearly the best, and I appreciated it. It was also true that no one did the homework. I would always review it anyway. This gave another opportunity for the students to see how to do certain problems before I gave them some in-class practice of their own.

Jack continued, "But you are wasting your time. For example, I like math and I'll do my best at school, but I'm not going to let anyone know I'm good at it—same with the rest of these kids. If they go home and do their math homework, or if someone finds out they are good in math or science, there will be an uncle or a father who is a Crip or a Blood, and he's gonna beat the shit out of the kid, because they don't want a nerd in the family."

Huh, so here you stand a science teacher, ready to teach. Your first assignment is likely to be in an urban middle or high school, and you will be the gatekeeper to Nerdsville, a place they don't want to be—where they fear to be!

Yes you are weird. You are strange. You are anti-cool. So don't try to be cool. You must find your own circuitous route to get the kids on your side. But that is going to take some time. Just don't try to be "all that" on day one, because you can't.

If you are ready to sign up to teach science, you are in for a lot of fun—and some pain. My goal for this chapter is to boost your fun and shrink your pain to nothing but a singularity—and to make you better than the 99%.

This chapter is a mini-version of the book. It's your emergency read, your Cliff's notes, your one-word post-it reminders you put on the side of your computer screen. And, amazingly, if you follow it you will indeed be better than 99% of your science teacher peers.

SCIENCE TEACHERS

Science teachers are a special breed. We are generally not as ebullient as say, a drama teacher. If you haven't taken the Myer's Briggs personality test, try an online version. They are all pretty good. Personally I test out as 100% introvert—seriously. My kids don't see me that way because I have evolved a persona that works for me, but on day one it was a problem. I wanted to teach my kids all about graphs, but they were break-dancing on the floor, with condoms spilling out of their pockets. In Eighth Grade.

For most of my life I tested specifically as an INTJ, one of the rarest of the 16 personality types, represented by only 2% of the population. We INTJs are independent, skeptical, and hold others and ourselves to high standards. This personality type is also known as "the Scientist". No surprise there. Eventually I morphed into the INTP personality, also rare, and sometimes known as "the nutty professor". We INTPs are "interested more in ideas than in social interaction". (from http://www.myersbriggs.org)

I am sure I represent a lot of science teachers. In fact, I am so anchored in these nerdy personality types, that by statistics alone, many science teachers are likely more human than I am. (Science and math comprise roughly 33% of the curriculum, whereas these nerdy profiles are just 5% of the population.)

Regardless, we science teachers and our misfit personalities are needed, but school might be a difficult fit for us. To survive, we need to adapt.

I have known a lot of science teachers. I've taught in eight schools and two districts. I've also been a science administrator and a vice principal. I've been through professional developments with science teachers. I have observed them. I have gone to pubs with them. So I know that what I am saying is largely true: many of us are not that animated. We physics teachers are the worst.

So what happens to science teachers? A lot of them leave the profession. Some that stay adapt by getting even more clammed up, more demanding and hard. I once saw a woman who did nothing but screech at her kids. She gave them a chemistry portfolio project that was to last all year. I never saw her effectively teach anything.

But science teachers can become the very best teachers at your school. They can lead by word and by example. These accomplished science teachers are often some students' favorite teachers. So there is hope.

It all starts with you knowing who you are, because it is then that you realize your students, being different, learn in different ways. This is an important step on the road to exceptional teaching. And you can take that step on Day One.

SMILE

Some teachers say they won't let a student see them smile until Thanksgiving. And they mean it. I've seen their classrooms, quiet and controlled—and oppressive. It's not what I'm after.

I think there is a lot of wisdom in using principles you will read in those classic books about winning friends and influencing people. In fact, Dale Carnegie's book is a good one.

No, the student is not your friend, but he could use a smile—not the kind of smile that says, "Hi there little man! Today I welcome you to walk all over me." Rather, use the smile that says, "Namaste, I see you."

Another thing Carnegie's book says is that the most precious word a person can hear is her name. Think about it, how many of your teachers regularly called you by your first name? And when they did, what did you think of that teacher?

So, I combine both the smile and name idea into a convenient package I use at the very beginning of the period. The time between classes is a time you can do things to get caught up, but I like to take a couple of minutes to stand by the door when they come in. Sometimes I'll say, "Hello Susan," or "Here comes Ray!" Sometimes they respond by rolling their eyes or by reaching out to shake my hand.

Sometimes I'll tinker or fidget by the door, but it still gives me a chance to connect.

Last week I noticed the pencil sharpener by the door had been broken off its steel mount. The metal was ripped clean off. So, I had fun with it, tying the sharpener to the base, making a pendulum out of it—hey, a physics tool *and* a conversation piece. As a bonus, if the custodian sees the sharpener pendulum, he might think to replace it.

RELAX

As a new teacher you have two main sources of stress. One is your thought that you have to impress the administration. The other is your thought that you have to control the kids. These two thoughts work together in a mad feedback loop, because you probably know that the clearest way to attract the administration's worst attention is to have no control over your classroom.

Stop. First of all, the administration needs you. They don't want you to go away. Second, the worst that could happen is you lose your job, and guess what? Most jobs are way easier than a teacher's first two years. So you can't lose!

Class control is the most important skill for you to master. Teacher education programs forget this because the professors who lead them are often accomplished teachers who gained the art of classroom management so long ago they forgot they were ever doing it. If instead teacher trainers were ex-principals, they would make classroom control the most important part of your education because they see the results every time they visit good and bad classrooms. In fact it is the number one criterion for judging good versus bad between classrooms. That's why I devote a chapter to this topic, and I give a few pointers here and in the next section.

Relax because your very first day or two is a gift. The students don't know what to expect of you. So they are watching. They are giving you respect. They are hoping you are a good teacher, and yes, maybe a few of them are hoping they can find a way to take advantage of you. But they are watching, and they are giving you some time. Good, take it. The most important thing for you to do is relax. Establish a pattern of being calm and controlled in your classroom no matter what.

Have you ever seen Cesar Millan's TV show, *The Dog Whisperer*? If so, you know that remaining calm is the most important step in gaining control over a rogue dog.

There is a simple primordial reason for this: not being calm implies weakness. Maybe someone is challenging your tribal leadership, and it's rattling you. Don't let anyone see you shake. You are the leader.

For the same reason and more, don't scream. Don't ever yell at a kid. Obviously you are not calm when you do so. Furthermore, you are not exercising strong leadership. Sure, it might make you feel powerful for a second, but the truth is you are yelling because you feel your power slipping. You are further eroding that power by yelling. Plus, you've erased all that good feeling your student had from the time you greeted him at the door.

If you can stay calm and not yell, you are off to a fantastic start!

LEAD THEM

I could have titled this section "manage the classroom," but that is boring. Businesses can be managed or led too. I'd rather work for a leader.

So a leader starts with a vision, a purpose, and a mission. I lay this out on day one mostly because I really want students to get to the promised land, but also because I want them to know I have a reason when I ask them to stop talking and focus.

Currently I am teaching regular physics (grades 9-12), International Baccalaureate (IB) Physics 1,2 (grade 11), and IB Physics 3,4 (grade 12). This is a great schedule because each of these classes lead into the other. That means I get to have some of the same students three years out of four. This rarely happens in schools, but when it does, it's great for the classroom. It is also convenient for me, because my goals can be identical at each level.

I start the year off by laying the syllabus under the document camera. I explain that physics is the foundational science. I might trace a standard's connections through the other sciences, for example, how electric forces keep molecules together, and how certain molecules make life possible, or give food its taste. Then I'll talk about how people in the 19th century felt electricity was in some mysterious way connected to and responsible for life, which is why Frankenstein's Monster was zapped with lightning. There is no need to go over each standard, just one good explanation on one standard implies all the standards have such relevance and importance.

Then I'll talk about the set of state standards, how they were chosen by a group of experts, and how this lone odd standard on *transistors* is there because some representative from Silicon Valley insisted on its inclusion. This honesty builds believability. But I emphasize my belief in this set of standards, and that I intend on dealing with all of them well.

Then I mention the IB standards, how they are broader and deeper and more fascinating.

Next, I will tell them one of my goals is that they acquire a power over the material world by their ability to apply their knowledge through concepts and computation.

I will follow this by a similar goal, that they acquire an appreciation for the beauty of the natural world, as seen through the eyes of physics.

Finally, I tell them that I want them to beat all my former students by a) getting higher average IB test scores as a class, b) getting a 100% pass rate, and c) exceeding the school record for number of perfect 7 scores on the course grade.

Students enthusiastically buy in to this last set of clear and measurable goals. The hardest goal to surpass is that last one, but if anyone can do it, it is my current set of incredible seniors. I have one young woman who has created a study guide for herself It's so good she should publish it. I believe having these goals helps motivate this kind of response.

None of these goals would work on middle school students however. They don't see school the same way college bound seniors do. You might want to hit on a few interesting highlights, like, "You will learn to identify six constellations. And, if you get lost at night, you will know which direction you are walking by keeping an eye on them."

Then you will go over your very short list of rules, which, you say, you need to "keep things flowing". My rule #1 is "No talking when I'm talking." To make it easy on them to follow this rule, and when I want their attention, I always tap some beakers with a meter stick. They sound like muted bells. Students know this is my sign to listen.

After you make your list of rules you need a set of consequences. I rarely have to do anything more than a) ask the class for silence, b) wait a couple of seconds as I stand calmly and look around, and possibly c) ask a specific student by name to get quiet.

If you are new, and especially if you are teaching at the middle school level, you will need to rely on systems that take you to a fourth, fifth, and sixth step, with different consequences at each step.

In my list above, only the third step had a consequence—students don't like to hear their name in a negative context, like being asked to not talk.

Your consequences should be tailored to your classroom, but here is one example of how to escalate consequences. Step four could be that the rule breaker stays after class (just a minute). They hate that because they use passing periods for socializing. Step five could be a phone call home. Step six could be a referral.

Using a system like this takes time to get right. You don't want to escalate things too quickly or it takes way too much of your time to enforce. And, you will lose all respect if everyone's getting referrals.

On the flip side, it is even more important that you are a) consistent, and b) fair. This means that if someone the class perceives to be the teacher's pet is allowed to chat away while someone else gets the hammer, again, you will lose a lot of respect, and things won't work right.

Such systems are best when they employ some reward, like an end-of-the-week celebration. For example, if there are no referrals all week, the last 15 minutes on Friday could be a game or free-time.

Make it easy for people to comply. Make two posters and post them in big letters side-by-side. Title one "Rules".

Title the other “Consequences”. If you don’t like the harsh tone of those titles, then call them “Expectations” and “Results”. Make sure to include *celebration* on the consequences/results poster.

At this point I strongly recommend you get a book on classroom management. Lee Canter’s *Assertive Discipline* is a classic, and it worked for me. It expands on what I am explaining here. Even many experienced teachers read this at the start of each school year.

It was no secret to anyone that in my first assignment at Weller Middle School my classes were out of control. I didn’t get much feedback, and I was incredibly tolerant of the madness around me, so things got really “spicy”.

It is bad when that happens because it is so hard to turn these situations around. This put me in a desperate and depressing situation. I had recently left the engineering position I had for five years. I took a 50% pay cut to try out this teaching thing. Meanwhile, my wife and I just had our first baby, and we were intent on her not working while the baby was young. I had to make this work.

So, I did everything I could: I went to workshops on classroom management, I talked to other teachers, and I read everything I could get my hands on. Then, like a true INTJ, I implemented a plan, following it scrupulously. It felt strange for me to ask, be consistent, and follow through. But I did. Miraculously, it worked.

By the middle of that first year, my room sounded like a library. My university teacher education program even started bringing teacher candidates into my room to show how it works.

Then, a soon as I was able, I transferred to a high school and was even happier! (Some people prefer middle school. Back then I did not.)

ENGAGE THEM

Lecture works, especially for some students. It can be their preferred way of getting new information. But it is too easy to think you can deliver concepts via the *blah, blah, blah*. Furthermore, it really doesn't work for more than a few minutes at a time. I've heard people say ten minutes, then switch to something else.

What is that something else?

- Videos
  - (especially short ones, not movies, but science concepts made visual. However, this is really much like lecture, a passive way to absorb information)
- Simulations (e.g., https://phet.colorado.edu/)
- Demonstrations
  - When the teacher exhibits a phenomenon that demonstrates a concept, (e.g., adding concentrated hydrochloric acid to a saturated salt solution so that crystals precipitate—the common ion effect, or making things explode—combustion reaction)
- Labs
  - A physical investigation into some concept or relationship, followed by a write up.
- Mini-Labs
  - A quick physical application of a concept, assigned in conjunction with the completion of a chart or graph or summary paragraph
- Worksheets of problems
- Projects

    - An in-depth lab with a self-designed component, or an extended activity with directions to build an apparatus that behaves competitively with others' devices, can be graded lab-style or from a rubric focusing on certain report features
- Tests
- Quizzes
- Verbal quizzes
- Small group study times using pre-tests
- More lecture with turn-and-talk mixed in
    - Turn and talk allows students a minute or thirty seconds to explain an idea to a neighbor for the purpose of keeping the class caught up and to drive the learning deeper. It also gives students a chance to re-set and take in more lecture.

And many other possibilities

Engagement implies far more than *modes of doing*. Engagement at its best elicits students' passions to explore, think, and do. You will find details in the chapter on motivation. Fun is also an effective element of engagement. See the next section.

BE FUN

What makes something fun? It does depend on the person. Your idea of fun might not be the students' idea of fun, but you will figure that out over time. When in doubt, blow something up.

It doesn't really matter, because you, being the leader, need to have fun. So, do what makes sense to you, at least for starters.

Some people are not natural clowns and that's ok. You don't have to be a comedian, but there is a way to make students enjoy what they are doing.

A few years back I taught at a brand new school, Gigantic High. I had just unloaded a bunch of boxes of equipment and supplies. They were stacked up in the corner of the lab off to my right. The stack was tall and deep. I was about to give a verbal quiz. I told the students the first two people to give me a correct answer would get to attack the box stack, MMA style, "like this".

I ran headlong into the stack of boxes and thrashed around, legs and arms smacking boxes as I sunk out of sight.

When I got back out and looked at the class, all the boys' faces were lit up like they just got a new Xbox on Christmas morning. They were half standing up, ready to go. The girls on the other hand, were horrified. So, I don't know if that was a net positive on that day, but I suspect it changed the dynamics for the better in the long run. Of course, I delivered on the MMA promise. Two or three boys got a turn.

Sometimes ideas follow you around for a couple of decades before you put them to use. This is one. My first year teaching at Weller Middle School, Mr. Pipetto asked me to help him open up a dormant classroom trailer. The back windows were nailed shut.

We saw a metal pole lying on the ground outside, so we took it into the room. We held it over our heads and charged straight at the plywood covering the window. The nails held, and the wood barely cracked as the pole bounced off.

We went back outside and found an identical pole, but this one was filled with concrete. It was much heavier. We took it inside and charged again. This time, when the pole hit the window, the pole kept right on moving. The nails slid cleanly out of the frame letting the plywood fly out the back. What a great demonstration of Newton's First Law—and I happened to be teaching it at the time! Unfortunately the kids weren't around to see. But seeing was not the dramatic thing for me. Feeling was. Having those two poles in my hands and feeling the power of that second one was compelling. I understood momentum and inertia in a deeper way after that incident.

Twenty years later (last year) I realized how to put that experience into kids' hands in a fun way: Smash Lab! I asked the staff to get me any discarded furniture they wouldn't mind destroying. We got several little pieces, but our art teacher gave me a whole couch!

Luckily, I have a huge blacktop patio just outside "the dungeon," my classroom. That's where I stacked the pieces. I brought in a sledgehammer from home. I took volunteers, first giving them a stick and some safety goggles. I told them to smash. Of course, not much happened with just a stick. Then I gave them the sledge. Destruction followed.

Since I had three other sections of physics, not everyone could get a turn, but everyone watched. In addition, I had a few smash-labbers tell the class what it felt like to change from the stick to the hammer. We summed it up with a discussion of mass, momentum, and inertia.

Furthermore, "Smash Lab" was just a fun thing to say. We got to say it on the days leading up to it, and we got to say it again every time we referred back to it. "Smash, smash SMASH!"

My final example, for now, is a lemonade-out-of-lemons story. Just a few days ago, Period one started with another appearance of Mr. Roach. High-school-aged students are mostly terrified of roaches. So it was that morning. Jose was a huge kid, and one of the smartest in the class, but he was out of his seat in a panic as Mr. Roach scrambled underfoot. The whole front of the class was screeching.

This has happened occasionally throughout my career, but ever since I moved into the dungeon, I get a roach attack about once a month. I'm sick of it. It destroys instructional time. I have to get rid of the roach; I can't ignore it. Then the students have to calm down.

So this time I decided to go with the flow and at least have fun with it. Plus, if the students see that I'm not afraid, it should be quicker to calm them down afterward. So, I marched straight at Mr. Roach, grabbed him with my bare hands and threw him at the Promethean screen.

That was a surreal move. Mr. Roach was back-lit by the screen, so you could see every spiky detail of his silhouette as he fell. I noticed that he seemed to have the same startle reflex as an infant, spreading his legs out to brace his fall. Of course, I couldn't leave it there. He still had to be caught and dealt with.

I grabbed an overflow cup, scooped him up and put him under the document camera. I kept zooming in until the class could see every move of his joints and antennae. Under magnification he looked like some sort of spastic robot. The kids expressed repulsion. Soon one of them shouted, "Eat it Burtson!" But there would be no geek show that day. I let Mr. Roach go out the back door.

Interestingly, students must have done some talking, because later that day other classes asked about what happened with the roach. I think they had fun. I know I did.

If you don't see yourself doing the MMA Box Dive, Smash Lab, or Roach Launch, there are less aggressive ways to be fun. You'll find some of these in the Fun chapter.

BE BIGGER THAN LIFE

I'm not much for decoration, and I realize my classroom can be drab. So, early in my teaching career I sought out people who had special décor in their classroom, and I observed what they did and how they taught. There was this one biology teacher I remember who dressed 60's-psychedelic. He had hair like the Monkees, and his classroom looked like it was done-over by Andy Warhol. He had rafters in his room. He strung long streamers of kelp and seaweed through them—straight from the beach. I asked how long he kept the weeds up. He said just until they started to smell.

I still don't decorate so much. I wish I had the knack, or took the time to get someone to help me, but I have found a different way to be bigger than life. See the *persona* chapter.

After I left Weller, I got my first high school job teaching chemistry at Heinlein High. I went through a stage where I dressed like a Mercury-era mission control geek. Other than Halloween, that's about all the dress experimentation I did. I'm sure I was influenced by the famous Dr. Stella Groves.

Dr. Groves was an honest to goodness NASA engineer from the early days of the space program. If you see the movie *Apollo 13*, there's nothing but white men in the control room. That's not true to life. Dr. Groves was there back in the day. She calculated trajectories—in a mini-skirt. Then she came to the district to teach physics.

Stella was near retirement when I became a young science teacher, and I just loved her. She wore that mini-skirt all the way through her career. I think it was her statement that she could be a girl *and* kick ass. That's bigger than life (well, maybe it is life?)

Being bigger than life means more than dressing up. There was a glorious time in my career when our entire science department at Sakura High just resonated with energy. We started a grass-roots observation cycle, looking at our buddies teach their kids so we could give feedback and get better at what we did.

Dr. Savanna Rocco was one of the perpetrators of this action. She was also one of those "best teachers in the school" I referred to above. One day I was teaching about *beat frequency*, how slightly-out-of-phase waves give a slow reinforcement/destruction cycle as they interfere with one another.

I thought to illustrate this with two lice combs. They had small spaced teeth, but one was slightly tighter than the other. I lay them on top of each other on the overhead projector. The result was a regular pattern of dark and light regions of interference, a perfect demonstration.

Savannah was in the room, watching. Afterward, she laid into me: "That was so awesome! What a great demonstration! And you were so deadpan. That's *not* how you do this!"

Well, she was right. I love science, and I thought this demo was pretty phenomenal. But I didn't act like it. I realized my attitude was, "These kids are spoiled, they won't get it, but I gotta show this anyway, maybe someone will think it's interesting." Well, I AM THAT SOMEONE. And I have to be that someone. Again, that's leadership. They won't believe it if I don't show that I believe it.

So now, things are different. "OMG! Did you see that?! What was that about?" I get it. And it gets me excited. Now, I won't stop at the lice combs. Being animated keeps me ranting. I'll make more connections. "Do you ever see two fences in a field when you are driving, and one is in front of the other going off at an angle—you see this same pattern! And, is there a musician in the room? How do you tune your guitar? What do you hear?" And sometimes I'll bring in my own guitar and play the beat frequency. Or, I'll take two identical tuning forks and heat one up so it is slightly off, and I'll get it that way.

Anyway, my message is that when you teach you represent something, and if you play it up, it can be to your advantage. When you start off in your career, you deliver lessons, and it feels kind of strange, you don't know how to act, how fast to go, how much of yourself to put into it. Once you figure that out, you start catching tread. Soon you will have arrived at a place you can build on. See the chapter on *Persona* to get the full explanation.

BE PRESENT

You are probably thinking I'm telling you to meditate or be mindful. Naw. I do that and it probably deserves its own chapter, and maybe it will make it in the second edition.

All I'm saying is that you should participate in school life. The returns are real. First, it's great to see your kids in a different context. I had one very quiet girl this year. I never heard her speak above 60 dB, but on the soccer field she was the most aggressive offensive player. She got an award for it.

Second, and more importantly, kids appreciate seeing you at their events. You don't have to stay for all four quarters, three periods, or two halves. It is hard for me because I have some side interests going. When I was younger, it was hard because we had little ones at home. But I'll make it to a few things a year, and it's usually the most fun part of my day. You see the kids playing. You see some of your other students in the stands, sometimes with face-paint on, going crazy. Third, you will likely run into other teachers and get a chance to get caught up or just joke around.

If you are a new teacher, it is especially hard to go, since you will probably need time to prep. However, you will make quite an impression by showing up to a few events.

The next level up from going is to actually be a coach or a club advisor. Talk about classroom management—coaches have no problem with that. You become a coach and you are an instant kingpin on campus. But I don't see how a new science teacher, or any science teacher for that matter, could have time to both teach and coach well. Science teaching takes a lot of time to do right. So does coaching. By the way, if a student ever slips up and calls you "coach," that's a huge compliment.

Club advising is a great alternative. Our school has an Academic League team. I share the sponsorship of that club with a friend. So many of my physics students are on that team that I often comment on the previous night's match during class.

I HATE BILL NYE, BUT KIDS LOVE HIM

I have always had a certain repulsion for Bill Nye's image. When you see him talking, there are often flasks in the background, filled with colored water and rising bubbles. When I think of him I think of "green goo". There's too much goofball in his presentation. I prefer profundity.

Like Neil deGrasse Tyson! Tyson is almost spiritual at times. Look at his "Most Astounding Fact" at https://www.youtube.com/watch?v=9D05ej8u-gU.

But the truth is, kids love both these guys. Both viewpoints are needed, and many more. In fact, Bill's seems to saturate our culture more broadly. I once went to a big science teaching convention in another city, and there was a huge line of teachers waiting for an autograph from Bill Nye. Maybe my original distaste for Bill Nye was because I was somewhat like him, and didn't want to admit it. Yes, I am a bit of a goofball myself.

LET THEM KNOW A LITTLE SOMETHING ABOUT YOU

Every year I bring in my tortoise at least a couple of times. Griffin is my random number generator. I place him at the origin of Cartesian axes I chalk into the asphalt. And I let him go. He helps me teach *speed*, *velocity* and *vectors*.

Students love Griffin. They feed him. They ask when I'm going to bring him next. I explain the story of how my daughter wanted a turtle, and a week later he walked up my friend's driveway eating flowers. Sometimes they think he's always around, just sleeping in my equipment stockroom.

In college, and during my first few years as an engineer, I was in a band called aMiniature. It was more than a side project—we really wanted to "make it". In our early days we played clubs like *The Spirit*, venues that welcomed new bands. Many times we played to no one but the waitresses. Then, we started to get some followers. Eventually we headlined shows. We played big local venues like the *Casbah*. After five years, I finally quit the band to get on with my life. They went on to get a record contract and put out several albums. As I write this, they are preparing a reunion show. I am looking forward to seeing it.

My time struggling with the band, trying to make something new, working out problems with people and each other—it is all more than a memory; it influenced how I am today. But I never thought to share any of this with my students.

Then one day in my fifth year teaching, I was in my IB Physics class demonstrating beat frequency with my bass guitar. One student asked, "Mr. Burtson, were you ever in a band?" I said, "yeah". "What was your band's name?" (I could sense a little sarcasm.) "*aMiniature*. We had to put an *a* in front because there was a jazz trio in Chicago called *Miniature*." Then, no questions, the class fell silent. They looked at each other in disbelief. It was a small class, only eleven people, but half of them knew the band. A few of they guys closely followed two local bands, *No Knife* and *aMiniature*. Some of them even had email passwords using *aMiniature* song titles! Five years in and I never mentioned this, and they might never had known, but for a random question.

The advantage of them knowing this about me is instant legitimacy. If I participated in something they valued so much, maybe what I was doing now was important too? It's like the coach status I mentioned above. In fact, any mentioned hobbies, from knitting to coin collecting could build connection with someone.

Obviously, you are taking a chance by mentioning your place of worship or some common interest in any public figure who attempts to "break the internet." And sure, I might have thought my participation in indie/alternative music might alienate me from a death-metal or a Hip-Hop kid. This is why I don't wear Doc Martens to school. But something like this is worth mentioning.

Did it help with my relationships with these students? I think so. It's hard to say, since they were already so motivated. However, I have other evidence.

Last year I was signing in for the day. Standing in the admin assistance's office I just spoke out loud, "I'm starting a Joy Division Tribute band, and I'm looking for members." My buddy Todd, a law teacher, heard me. He said, "Are you serious? Because, I love Joy Division. I can play guitar and I can sing." And so, Old Order was born.

Todd, my son Alex, Dan, a drummer friend from another college band, and I practiced that summer. Come fall, I did not keep it a secret that we had an all-ages show booked for the Che Café. About 10 of my students showed up, moshing there in the dark in a quaint wooden building in the forest at the edge of the UCSD campus. What a great time that was. But it doesn't stop there.

We continued to get shows around town. A parent group was holding a fund-raiser to make money for the school's band program. They found out that the students liked Old Order, so they asked if we would auction off a performance of our band at the fundraiser. Of course we agreed.

Our house party package attracted the highest bids, so they asked if we would do two shows. We agreed. Later, one homeowner got cold feet and donated the performance to the school. The parents group parlayed that up to yet a bigger fundraiser. There were food trucks, lights, a DJ and a student band playing with us. To top it off, the drum-line learned two Joy Division songs and played along with us. The "Dead Souls" beat was ridiculously fun. Mr. D's drum-line students kept the beat they worked out and used it for regular competitions.

So what's the moral? It's a classic win-win story. More importantly, the connections Todd and I have with the parents, the administration, the band director, and the students are worth far more than the money we made for the school.

As I write this, a student photographer is printing off pictures of the show, which I will display in my classroom. My favorite two are not of the band, but of our kids moshing and of our students posing for a picture, everyone with giant smiles.

That's Chapter One, a mini book in itself. Here's a summary:

- **Know thyself**—Accept your image as a science nerd and use it in your favor. Know that students learn in different ways.
- **Smile**—Don't be a black hole, be human!
- **Lead them**—Give them reason for what they are doing. Give them something they can work together with you to achieve. Tell them what you want; address mistakes with fairness and consistency. Celebrate successes.
- **Engage them**—Mix up your instructional modes. Have interesting things to do.
- **Be fun**—both in manner and in how you teach.
- **Be bigger than life**—Obtain a persona.
- **Be present**—Participate in school events.
- **Don't hate Bill Nye**—Understand that different science personalities resonate with students. Find out how you can be one.
- **Let them know a little something about you**—hobbies, interests, and achievements. There are lots of students and one of you; the more you share, the more connections you make.

You will note I said nothing about lesson planning here. If this truly is your first day, you will have a remarkable lesson plan, probably whipped up as an assignment in your teacher education program. Fine. Use it. It's going to be great. But everything in this chapter is really what you want to do.

2. Toolbox

MARKERS AND PENS

Towards the middle of my first week teaching it became time to grade something. I looked around, but I had no red pen. How come no one gave me a red pen?!

Once I opened my door to the neighboring classroom, the rest of the walls were all partitions. There weren't even doors between rooms, just gaps to breeze through.

It was late in the afternoon. I was the only one left around, so I just drifted from teacher's desk to teacher's desk looking for that red pen. Finally, there it was, on a math teacher's desk. I resolved to return it, as soon as I tracked down my own red pen.

Some teachers don't like to grade with red pen. "It's too harsh," they say. "It makes them think I'm angry." They use purple instead. Fine, go down to the teacher's supply store and buy yourself a purple pen. I would rather use a standard issue medium point red pen. It gets the message across.

So who has these red pens and the rest of the stuff you'll need to teach? Does the principal keep them in her desk? Does the department chair hoard them? Are they in a pen dispenser in the teacher lounge? No. *The head custodian is your gate-keeper. Be nice to him.*

The custodian resides at some edge of the campus, close to driveways that feed into the main access road to your school, often near or connected to the main office. At a high school their headquarters might be next to the gym or library. Look for electric carts zipping to and fro. Often they will lead you straight to the custodian's bat cave.

When you see it, go in. The best time of day to find the head custodian there is within a half hour of the start of the school day. Ask him for what you want, and he'll get it for you.

For you financial wizards, you know a recession is just around the corner when schools get loose on their supplies policy. So, if you show up to ask for a red pen, and no one's around but the door is open to the supply shelves, go get your pen, but be warned: the market it about to crash, and the state tax revenues are about to narrow down to a trickle. You'd better get two red pens.

In fact, to start your first day, ask for some red, black, and blue pens. If you have white boards in your room, ask for some dry-erase markers, a multi-color pack. Ask for some regular markers too. If your school is rolling in dough, ask for 12 packs of them, for project poster work. Ask anyway.

Finally, snag some vis-à-vis fine point markers. I'll tell you why later. Oh, and get a highlighter. This is all you need for day one, but I'll tell you what else to get in the next section.

MISCELLANEOUS

The pen is mightier than the sword. That's why I gave it its own section. Now that you have the various forms of writing, the rest of what you need to operate comes in this section. From least important to most important, here you go:

- **Rulers.** You measure stuff in science. Get a class set of rulers. First ask the custodian. If that doesn't work out, see the catalogues section below. Rulers are cheap.

- **Protractors.** See Rulers.

- **Mic.** When our district re-tooled to be high-tech, they gave us microphones. I have been trying to get Todd, my singer, to use his mic to preserve his voice, but he hasn't yet embraced it. Too bad. The minute I started using my mic, my voice changed from aggressive sage to improv comedian, and it made everything go smoother, "Joe Schmoe to the front desk, Joe Schmoe to the front desk please."

- **Calculator**. Just how INTJ are you? Remember Ray Man? He didn't need a calculator.

- **Seating Chart: Acetates and Post-its.** Now we are getting serious. One of my most useful tools is this combo. Wouldn't it be nice to be able to glance around your classroom, see who's absent, write it down and get on with teaching? Here is an ancient tool that blows away any technological attempt to trump it. What you do is. . .

    - Create an outline for your classroom on paper. Using a ruler, sketch the desk positions. Photocopy 6 periods worth of this graphic (or eight if your schools is that crazy). Then take the smallest post-it notes, cut them in half, and place them at each position.

    - After you have established who sits where, alphabetically, randomly, or preferentially, etc. pass around the chart and have people print first and last names clearly on the corresponding post-its.

    - When you get it back, tape a clear acetate page over one edge (your custodians have these acetates). You can write on these with your vis-à-vis pens! "A" is for absent. "T" is for tardy. The markings wipe off with a damp cloth or napkin.

- But wait, there's more! If I am giving a verbal quiz or some other spontaneous assessment, I can write that score down too, and easily record it in the gradebook later. This is magnitudes easier than collecting papers and grading them, though you WILL do plenty of that too.

I also keep a tally of homeworks completed over each seating chart name. I make homework (HW) worth anywhere from 10-20% of the overall grade. But I don't record every day, and I DO NOT grade each HW assignment. Instead, when students are doing the warm-up activity, I walk through the classroom and write down a tally mark on each name that did not do the HW that day. After ten assignments or so, I enter a grade. Seven out of ten is 80%. How? One freebie. One-hundred times (Ten-three+one)/ten = 80%. The chart is also helpful when you have to take the students outside on a fire drill and you need to make an accounting of who is with you.

The reason you used the post-its is so you could easily move the seats. Two people talk ALL THE TIME? Then move them to opposing corners of the classroom.

- **Beaker, Bell, Chimes and Guns.** During that first year of teaching, when I was trying to dig myself out of that horrible hole of nightmare classroom behavior, I went to a workshop that showed me how to use a gun. It was a fluorescent toy that made wacko squealy sounds when you pulled the trigger.

The presenter said that the sound becomes your way of communicating to students that you are transitioning to another activity and/or you need them to stop talking and listen for instruction. I use this strategy to this day. It is essential.

I stopped using the gun when my vice principal called me to his office. “Mr. Burtson,” he said. “I hear you are using a gun in your classroom.”

“Yes, it’s really effective! I learned about it at this workshop on classroom mana. . “

“Please stop using the gun.”

“But it’s just this green fluorescent. . .”

“I’M ASKING YOU TO CEASE AND DESIST. . .”

“Wow. Ok.”

This VP is an example of an extreme black-and-white thinker. “Guns bad!” It is my experience that you do the thing they are asking. They usually get hung up on some stupid aspect to a sub-point to a subordinate clause to a matter of policy. So, obey and step over like a grain of sand on the beach.

When I worked for a different district they issued a small block of chimes to every new hire. I used it religiously. It did the trick.

Currently, I use a series of four different sized beakers and a wooden meter stick. I often mix it up by smacking out some atonal melody to get their attention. Kids got so used to it that once, a student with a side DJ business showed the class how you could sample the sound the beaker makes, equalize its envelope of frequencies, and have it ring like a bell.

But almost anything would work: a kazoo, a series of claps, or a squealy fluorescent toy.

- **Meter Sticks.** See Protractors.

- **Measuring Tape.** See Meter Sticks.

- **Dice.** Dice are fair. Back when we did grassroots peer observations, we looked for whether or not we were favoring particular ethnicities or genders in our questioning patterns. With dice, it's all random.

I use two dice. One represents the row, the other the kid. Yes, it means I have a room of six rows, six seats across, but that seems natural to me anyway. Clusters of tables to make groups—those are evil to me. I can't get kids to stop talking in that scenario.

If I roll a 3 and a 6, that means that the kid in the third row, last seat gets the question. Don't accept "I don't know" as an answer. It really means, "I don't want to try." You can help them save face, if they were really not paying attention, but don't be too easy on them. Feed them some clues, but no spoon-feeding.

Some teachers, especially at the elementary level, keep a jar of sticks with students' names on them, and they randomly draw from the jar. Same idea. Other people write kids' names on a deck of cards.

I don't rely exclusively on the dice. At the beginning of the year, I tell anyone that if they want to sit in the front row, that means they are exposing themselves to unmitigated questioning whenever I want. I call this "the genius row". I feel okay about calling it that because they choose to be there! They are not necessarily the smartest kids in the class, but you gotta love how they promote themselves and welcome the risk. Finally, sometimes I just call a student's name because I feel like it.

- **Grading Program.** When I first started teaching I thought I was way ahead of the pack for using Excel to figure my grades. In fact, I was. Many teachers were still using physical gradebooks. It's easier to describe grading programs than those pencil-marked ledgers.

Doubtless your school will be using a grading program. Likely the district has a subscription. My favorite programs allow you to drop assignments and give special weighting to others. Most allow you to grade by points or by percent.

I don't get why people wouldn't use percents. I guess it could motivate a kid to think, "I'm only 40 points away from a thousand points in Mr. Smith's class, and I will have an A!"

I use this scale, which you can set in your program:

Labs, 35%
Tests, 30%
Quizzes, 15%
Homework, 10%
In-class assignments, 10%

- **Paper.** As we ascend the hierarchy of most important supplies, paper nears the top of the list. You will know why if you run out. You can usually get as much as you want, but sometimes there is a paper famine. We are currently in one. Seven years after our drop into the Great Recession, and schools are still feeling the pain.

Paper is one of those commodities that people start to hoard when things get scarce. I suspect this phenomenon sped up the dearth for us. A couple of weeks ago I was flipping through the cabinets and found almost a full ream of 8 ½ by 14 inch! I borrowed the paper cutter and chopped it all down to 8 ½ by 11. I survived on this for a couple of days.

Then, the day before spring break, one of my star IB Physics students was wheeling around a cart of paper packs for teachers! I still remember the white wrapper and plain black block letters. Yum. I emailed his dad to thank him, but he said it was his son's idea.

- **Science Equipment and Catalogues.** When you are a new teacher, it is a mystery what you need. In practice, I've found you use what you have. Sometimes you have to slip into what I call "toothpicks and Styrofoam" mode.

I strongly advise against buying things with your own money. This out of pocket budget can grow like a cancer. Count it a challenge to pay nothing for your science supplies. You are not supposed to pay anyway.

I once knew a nun who was a superb science teacher. One year she spent $25,000 of her pay to get her room set up with the latest computers.

In my first year teaching chemistry I needed to get a handle on what I had. I spent a couple of days the summer before doing nothing but taking inventory on chemicals. It gave me a sense of confidence to know what my chess pieces were.

When I moved to another high school to teach physics, I was horrified to find equipment that looked like it came out of the 19th Century. My thermal expansion apparatus was made of wood!

If it were the present-day me, rather than worry about an offensive piece of junk, I would find something else to do than measure thermal expansion in the lab.

Usually, there is a stream of money you can divert towards your want list: lasers, blocks of glass for index of refraction measurements, frictionless carts, etc. Different districts and different schools have different budgets.

When I started teaching, I spoke to the veterans who were around in the sixties. They said science teachers in California used to get thousands of dollars a year to spend on supplies.

My experience is more like $300 to $600. For the past five years we have gotten $0. However, I believe this will change someday. In the meantime, it has taught me to be frugal. You can build a pendulum out of almost anything! Think “toothpicks and Styrofoam” and cardboard too.

When the cash comes, you will likely buy from a catalogue of science supply companies. The district will have price deals with them. See your department chair for these catalogues, and ask if the order can be placed on the Internet.

- **Wall of Formulas.** Moses had his Ten Commandments, Hammurabi had his Code, and you have the formula wall! Luckily, I have a whole wall that can take pushpins. As we progress through the curriculum, I write relevant equations (and a few concepts) on paper and post them. They stay up all year. Students love the formula wall.

I do not see that it is important to memorize much in physics. The glory is in the application. Besides, they end up remembering crazy things anyway, like the charge of an electron and the mass of the planet.

- **Miscellaneous Miscellaneous.** You will quickly find out what is important to the disciplines you teach. Biology teachers love blood pressure cuffs. Chemistry teachers love acids and bases. You can get them a jug of muriatic acid at Home Depot. It comes in a green bottle and would go over well at Christmas. Physics teachers used to love stopwatches, but now almost every kid has a timer on his smart phone.

You will find out those essentials. Talk to teachers. Build on what you have. I am probably teaching physics today because of the monkey-and-hunter demo my high school physics teacher used. So, that piece of equipment is essential for me.

In summary, get these things:

- **Markers and pens**
- **Rulers**
- **Protractors**
- **A Microphone**
- **A Calculator**
- **A seating chart made of acetates and post-its**
- **A Noise-making attention getting device**
- **Meter sticks**
- **Measuring tape**
- **Dice**
- **A Grading program**
- **Paper**
- **Science equipment and catalogues**
- **A Wall of formulas**
- **Equipment that is essential to your discipline** (See Chapter 14: Lab Equipment and Supplies.)

3. Fun

Notice that this chapter comes two steps ahead of *Educational Theory*. Why? Because if no one's having fun, theory shmeery. In the last chapter I mentioned that I probably would not be a physics teacher today if it weren't for the Monkey-and-Hunter demonstration I saw in high school. Who knows? I might not have pursued engineering either. Maybe I would be a Star Trek Nerd lying on my mom's couch to this day, if it weren't for that Monkey-and-Hunter demo. And really, it was because the demo was fun for me.

The way it works is this, the "hunter" pulls back on a rubber film to launch a steel ball-bearing. The ball breaks an electromagnetic hold on the "monkey," hanging from the rafters half way back in the room. The ball starts flying towards the monkey at the same instant the monkey starts to fall. But no matter how hard you launch the ball, it hits the monkey. It could happen just as they are both about to hit the floor.

This demo shows the complexity of the effect of gravity on various objects regardless of their motion. It is a connection that is a sleeper on paper, but the way it plays out in the real world is phenomenal. It shocked me.

That, plus the clunking sound of the hit, plus the feathers and the goofy grin of the monkey all made it fun.

In the first chapter I gave a quick view of fun. One of my three examples was connected to instruction. The others were just fun mindsets. Fun plus academics is the ultimate, but fun is always welcome in a class, as long as it's not horseplay during lab time. Fun puts you in an open and receptive state of mind.

One person's idea of fun is different from another's, so I advise mixing it up. I promised to give examples of less aggressive forms of fun. Here is one. *Arsenal*, a biology teacher friend of mine, does a scavenger hunt on campus, awarding points to students who find objects from a list of clues. He also does a species identification activity in a local park.

Similarly, I ditch the traditional ripple tank and take my students to a local park to let them experience wave *reflection*, *diffraction*, and *refraction* in several different fountains. Walking somewhere in a beautiful environment can be more fun than standing under an incandescent light bulb. As a bonus, it makes up for not being able to afford those ripple tanks.

Fortunately, I was clued in to the importance of fun early in my teaching. That first high school where I was teaching chemistry had an interesting physics teacher. I think he was some sort of libertarian, definitely bigger than life.

He got to work when it was still pitch black outside. He would open up both classroom doors and get to it. His classroom was lined with those special gas burners that looked like upward-facing shower heads. He cranked them up full blast, making a circle of blue light around the room—just to get warm. He had a Vatican-style display of hair-band and spandex-metal posters hung way up on the walls.

One day I saw him drive a go-cart between buildings, coming from who knows where. I asked him why he had a go-cart. He explained the science of fun physics labs. I committed right then to getting my own go-cart.

So, soon after transferring to a school where I could teach physics, I found a grant application I could use to fund a cart. That's what I did, and kids loved my labs: Putting the pedal to the metal in *acceleration* labs, doing doughnuts in *centripetal force* labs, and slamming on the breaks in *friction forensics*.

But again, what's fun for someone else might not be worth it to you. I came back to my present school after the previous physics teacher left suddenly. People loved this guy, and they especially loved his *walk on water* activity, which had received some media exposure.

I did not like the idea. While I am comfortable letting my kids go crazy in a go-cart, I couldn't shake the image of someone doing the splits in wood-plank shoes, or smashing her head against the concrete side of the pool. Students and parents kept asking for *walk on water*, but they didn't find their Jesus in me. I had to let that go.

Smash Lab worked for me—and for the kids. That couch provided many opportunities to show inertia in action: arms falling off, back-rest cracking, and legs popping. But it would not go gently into that good night. It survived a couple of classes. Students were so committed to completing the task that I had to open it up to lunchtime destruction.

Interestingly, two of my IB physics students who very well might be among my all-time record-breakers—these were my most devoted smash-labbers. By the end of lunch they had reduced the couch to splinters and a billowing sheet.

Wow, everything I just mentioned is connected to instruction—yay me! But not everything has to be. Last fall it was punishingly hot in the dungeon. So, I went into the stockroom to tap into my giant bag of balloons (for static electric experiments). We filled them with water, went outside, and had a balloon fight. That's leadership, and it's fun.

Summary:

- **Make instruction fun.**
- **Make everything fun.**

4. Motivation and Engagement

Why do we get out of bed in the morning? Why do we do those things we are passionate about? And how do we get kids to take voracious bites out of our curriculum? Read on.

And if you want to dig deeper, I refer to another book: Deci & Flaste's *Why We Do What We Do: Understanding Self-Motivation*. Here's yet another book that has been popular recently, *Flow: The Psychology of Optimal Experience* by Mihaly Csikszentmihalyi. He's Hungarian, and I can't pronounce his name. He is a researcher, but lately has made his work public in a popular, applicable way. He also writes about creativity and happiness. All this connects to motivation.

You could also flip through my master's thesis: *"Pull! . . . Don't Push": Motivating Learning in Physics through Student Clubs and Projects."* I rely heavily on what Deci and Flaste teach. *Pull Don't Push* is available on Amazon. Or, you could just read this chapter, which will digest and expand on my thesis.

Why do you care about motivation? After all, you have a captive audience. If they want to graduate, they at least care about passing (well. . .at the high school level anyway). Simple. It's like rocketing into space. Contrary to public opinion, you do not have to achieve escape velocity to get to Mars. All you gotta do is keep your engines running. You will eventually get far enough away from earth's gravity so you can cruise the rest of the way.

However, you are going to need a spaceship the size of Montana to hold that much liquid hydrogen. Even then it's doubtful you'll make it. It's like going up a hill in a car. It takes gas just to maintain your position.

Instead, ramp it up to super-speeds, you'll do the job and maybe be the first to make it to Alpha Centauri. So this chapter is about ramping learning to super speeds by understanding human propulsion.

When you've engaged your students' motivation you'll start to hear things like, "This is awesome," "I love physics," and "I'm taking IB next year". If you don't engage them, you'll hear stuff like, "this blows," or "I'm gonna fail the test". I've heard both kinds of expressions, and I let them inform me.

As I begin to break it down, I want to apologize to my earlier self. I have another book out there about guys who break world records in the bench press. Naturally motivation applies in physical training too, so if I forget to quote accurately, I hope the author will forgive me. This book is *The Power Guru of Monsta Garage*. . .

TWO KINDS OF MOTIVATION

Schoolmarms from millennia past have refined methods to motivate kids: ruler smacks on the back of the hand, verbal humiliation, and dunce caps in the corner. This is called *extrinsic motivation*. At best, it makes kids work, for a while. At worst, it reduces their self-respect to the point they don't want to live any more. Extrinsic motivation sucks.

Intrinsic motivation comes from inside the kid. How could you know what makes a kid tick? Is this beyond your pay grade? Don't worry. It's mostly about you stepping out of the way. There are three spigots that open the floodgates of intrinsic motivation. They are 1) choice, 2) challenge, and 3) social connection.

CHOICE

“Choice” and “freedom” are battle cries you’ve heard before. If you love *Braveheart* as much as I do, you understand. Choice in science instruction means giving students latitude in how they proceed with their tasks.

Because I have seen the positive aspects of choice, I have grown to hate the typical lab procedure passed out by some science teachers. An instructor might think he is so on top of things if he prepares his chemicals and beakers and scoopulas and distributes a twenty-point bulleted procedure on how to do what it is he’s asking his students to do. My experience is that this converts future Einsteins into present day box-checkers. Yuk.

Here’s what to do instead. Give a task with a specified goal and a small set of parameters. That’s it. Don’t forget to address safety issues, especially if they are designing their own investigations. Approve these first.

I will give you three examples: a lab, a project, and an in-class assignment.

Lab. Ohm’s law is a central concept in physics. It’s also one of those neat, three variable relationships: $V=IR$. However, it is an oversimplification. IB knows this, so they want us to teach about internal resistance of a battery.

It takes energy just to use a battery, and that’s a side point, entropy rules in every context, even this one. So, I tell my students that they need to set up several circuits, take measurements of current, resistance, and voltage. Taking just one measurement might lull them into thinking they’ve nailed it, but the world is not that neat.

Multiple perspectives give various answers. So, what gives? Two circuits yield two equations, which is enough to solve for the unknown. If they make three circuits, they confirm or take an average. If they make four, they tighten it up even further. But it doesn't matter how they rig their wires, or really how many circuits they make. Their thoughtful choices give more accurate answers, which they can see in their multiple results.

In the end, I can congratulate them on what they've done, and I can acknowledge those who've done more work, because that extra work usually results in a better answer.

Had they used MY Voltage and MY resistance, they would have gotten MY answer, leaving no personal ownership over what they found, how they found it, or the effect of their choice or their persistence over the natural world.

Yay, they may have gotten the right answer by my method, but that leaves them Pavlov's dog, panting faithfully by the side of the cage, hoping and salivating for the next treat.

Project. *Thermodynamics* is often a lost topic in physics. Some people never get out of Mechanics/Motion/Energy. Within thermodynamics, *entropy* is often a lost standard. A thorough physics teacher might say, "Look at this child's bedroom, how neat and orderly." "And now, look how messy! Obviously his mom has not been around for a couple of days!" "That messiness is entropy!" And that's it.

I do an anti-entropy project with some simple constraints. I tell the students they are to build a device that when shaken, squeezed, or twisted results in less entropy (more order). Some students, often my "smartest" tell me that's impossible, that entropy must increase.

This shows that they have a rudimentary understanding of the universe. (See *Misconceptions* in the next chapter.) Yes, entropy rises and will eventually result in the heat death of the universe, but there are notable pockets of counter-intuitive results along the way, like the evolution of humans.

And I give them choice. Often, when I present a project, I give them examples of things students have done in previous years, but I refrain from doing that on this project. I want students to think deeply and broadly. I want to be surprised with what they come up with.

The first year I did this, I had a couple of student groups come up with the following device. They cut a two-liter bottle in half in the middle. They installed little plastic ledges at various heights along the wall. Then they hot-glued the bottle back together with flakes of paper inside.

When they shook the bottle, paper concentrated on these ledges—a perfect example of energy added to a closed system resulting in *decreased* entropy! And yet there are wildly different ways to show this.

Here are the parameters I used this year:

1. Make the device.
2. Demonstrate it.
3. Make an annotated sketch of the device.
4. Include a paragraph explaining a) how the device decreases entropy when you work the system, and b) how the entropy in the universe *increases* even though you've *decreased* the entropy of your system.

Note the broad brush-strokes. There is no detail, just a bunch of opportunity, which lends itself to personal choice, a cornerstone of motivation.

Below you will find a picture of an anti-entropy device before shaking, and another after. When the students presented this, they shook without looking at the box, kind of like how DNA brings order out of chaos, even though we don't have to watch it do its thing!

In-class assignment. This year IB made a nuclear decay simulation mandatory. So, it must be an important concept. Typically, teachers have students start with a bunch of pennies, ascribe a time period to each shake/dump cycle, and have the students select out the heads for counting. Pennies showing tails get tossed to the side as spent products of nuclear decay.

When they graph heads remaining versus time, they get a nice graph that shows the half-life, which is simply the period of time they ascribe to a shake in the first place.

So, how do we infuse more choice into this process? I start by giving each student pair one die and some graph paper. Then I put these instructions on the screen.

- Name some pretend isotope. I call mine "Burtsonium".
- Select a starting mass for graphing, like 2000g.
- Choose some time period represented by each roll, for example, 3.0s, 100 min, or 5700 yr.
- Select some decimal fraction multiplier between 0.70 and 0.95
- Make up a logical condition like, "multiply the current mass by the multiplier only if the numbers 1, 2, 3, or 4 come up on the die."
- Make a T-chart of Mass(g) vs. Time(s,min,yr), roll the die, and fill the table.
- Graph what you have.
- Use the graph to find your isotope's half-life.
- Make a paragraph explaining why this process successfully results in a half-life.

While the above might be starting to look like the punishing list I mentioned, it is filled with choice. Look at the verbs, "Name," "Select," "Choose," "Select, and "Make up."

Furthermore, they aren't told how to find the half-life, just to find it. Finally, they can say whatever they want in their paragraph, but they have to find a mathematical connection to this natural process.

When something like this works well, it's fun for me. I just did this yesterday, and I heard students say to each other, "My half-life was four thousandths of a second!" and "My half-life was two-million years!" Now I get to take out the CRC Handbook and show the students the range of actual values half-lives can have, and that they are even more extreme than they found.

These are a class full of intelligent juniors. To get them that enthusiastic, something has to tap their motivation. Giving them choice was one way to do it.

CHALLENGE

There is a lot to cover in the school year. I am proud of myself when I get through each standard with an array of opportunities for students to wrap their heads around. But to get the job done, sometimes the array becomes a single event.

Furthermore, it is tempting just to slap down a worksheet that gives access to the least common denominator—that person who struggles but you don't want to leave out.

The solution can be *challenge*. Sometimes challenge is a natural way to *differentiate* curriculum, to teach in such a way that each student is engaged at her level of optimal learning. To do this, *challenge* is best applied in combination with *choice*, see above.

Here is a simple example. Another required lab is projectile motion. One ubiquitous form of this lab has students roll a steel ball bearing off a table and into a cup. The challenge is to get your measurements and calculations figured out before you let the ball leave the ledge, so that when you finally do, you can celebrate a direct hit. And kids do celebrate a hit.

This year, I spiced it up even more. I knew there were some smarties who could use a little more challenge, so I told them they could use the long tracks and roll the ball off at an angle. Though that tweak considerably hiked up the intensity of the calculations, and though the curriculum did not require anything beyond a horizontal launch, I thought, "What the heck? If someone wants to do it this way it's more practice with vectors, which helps them understand physics and math better."

I sensed even greater satisfaction from those who hit their target this way. But I didn't stop there. A couple of my most capable students (the couch vaporizers) wanted to be even more exact, so we discussed rotational energy, which is another layer of complexity, and something that was removed from the curriculum years earlier. Of course, they incorporated this idea too.

You can also introduce challenge competitively, which can be even more motivational, as it pulls in a social element (see the next section). One example is my ice cube thermodynamics challenge.

I give students a few grams of ice. To be fair, they need about the same amount in each group. Then I have them get one large, one medium, and one small beaker. I tell them they can use any book in the class (for insulation), and one piece of paper.
The challenge is to arrange these components so they result in the highest percentage of ice remaining after 20 minutes of melting time.

The top two groups get extra credit points, though Deci would argue this is not a necessary step, since students want to win for winning's sake.

Next, they have to write a paragraph on how their arrangement was designed to keep out the most heat. I encourage them to use the words, "insulation," "conduction," "radiation," and "convection".

The next day, I show the best and worst functioning designs and we reason to cause and effect. Finally, they include a paragraph describing why the best designs had the advantage they did.

It becomes obvious why the winners gained the opportunity to gloat: In their cleverness, they might have used the book to shield from radiation *and* to insulate. Maybe they wadded their paper up into a fluffy ball to stop convection *and* force slow conduction.

Regardless, the learning initiates with challenge, is saturated in choice, and peaks in community.

SOCIAL CONNECTION

A few years ago some security guards at my school told me that gangs were not necessarily bad. They didn't need to be about criminal activity. They were just ways for kids to be together and to belong to something—especially in a crazy world that doesn't give them a sense that they belong.

I'm not advocating we start a social experiment where we encourage students in cities and suburbs to coagulate into gangs. But I do recognize the need for affiliation. So does motivational research.

High school has traditionally been a place for students to find that affiliation. Think Friday night football games, the cheering, the lights, the chants and banners, the “we’re better than you” attitude that we encourage and even laugh about.

Think about the cliques that form. Think about fraternities and sororities in college, and the hazing rituals they go through.

In fact, while low on the totem pole of development, the “we’re better than you” attitude is a natural part of our growth. It doesn’t make sense to think that just being from the same neighborhood, from the same school, or in the same set of clothes should qualify you to be the best, but that’s how people think. So, let’s use it to everyone’s advantage!

In class I let people form their own lab groups. Sometimes I do this with stipulations. If they choose on their own, they are likely choosing their buddies. The smart thing to do if you are not so good at science is to join a group that you know can help you understand. But it is rare that that happens. Usually, the buds gravitate.

The advantage is that the social connection is already in place. Good. But sometimes I mix it up, especially in a challenge activity where groups will be ranked by results or where a few will get extra credit.

Some teachers shy away from gender competitions, but I don’t. Sometimes it’s girls vs. guys. I have all four grade levels in regular physics, so I might make the requirement that they form groups by grade level. Then I play up the whole Freshmen vs. Senior thing, like they do in pep rallies and class competitions.

Believe me, you might feel a little like a cult leader when you do this, because you can whip your people up into a frenzy when report-out time comes. But hey, it’s fun, it’s motivational, and it gets people to engage in the activity. Isn't learning what it's all about?

Summary:

Provide choice, challenge, and social connection.

5. Educational Theory

During that first year of teaching, part of my problem was that I bored my kids. I still held the belief that Eighth-Graders would come to class hungry to learn science divorced of any context other than my talking face, *blah, blah, blah*.

I noticed that the teacher next door had no problems with her kids. She really enjoyed them, and they loved being in her classroom. I realized I needed to talk to her.

Right away Helen revealed the secret of her success: *Project Based Learning*. When she walked me through her classroom, laminated dinosaurs popped out at me from her walls. The room was decorated with similar creations from other endeavors she presented to her students.

She explained what is now obvious: Kids learn when they are doing, and kids learn better when they can connect what they are doing to other experiences and to their interests. That's project-based learning, in a nutshell.

Helen even told me to back off on the homework in Eighth Grade. She only assigned ten homeworks each semester. But even these were project-like, for example, estimating the number of visible stars by counting them through a toilet paper core and extrapolating to cover the whole sky.

THE SCIENCE OF TEACHING AND LEARNING

Imagine you are a space alien. You just got to earth and you are hungry. Somehow you end up in a giant supermarket. Problem is, you don't know what to get. Everything is in a neat package, looking equally important and appealing. Furthermore, even if you could see the contents of the packages, you don't know what tastes good, or what's good for your alien self. Without an earthly guide, you might end up loading your cart with sardines and dishwasher soap.

I felt much like that alien the first time I looked through all the possible research areas in my Masters of Ed program. Fortunately, I did have good guides. The Deci book was just a casual recommendation from one of my professors, yet it steered me towards motivation.

What you have in this chapter is a boiled down paste—just the good stuff. *Good* comes two ways. The first are those educational theories I favor through years of practice and personal interest, so their inclusion might be flavored by my particular brand of cooking. The rest pass through the gauntlet of time, common practice, and research about the research.

I wouldn't worry so much about my feeding you strange fruit. By nature I am a maximizer. I am constantly on the lookout for getting the most bang for my buck. If nothing else, this can be a great starting place.

MOTIVATION

It's so important, it has its own chapter. See above.

PROJECT BASED LEARNING: PBL

My intro to this chapter led you to this concept. Like anything, the volume of literature grows and grows in this area.

Projects can last a day or weeks. My own three kids went to a school ravenously devoted to PBL. My three are now all college age. They report that it would have been better for them to have more traditional instruction blended in.

I'll start with what can go wrong with PBL: Leeches. That's the worst thing—kids who just want to coast in orbit around the blue giants, basking in the glory of those students who do most of the work.

Another thing that can go wrong is *content deficiency*. If you are all-in on PBL, you will give students adequate time to luxuriate into a grand exposition of their findings. You encourage a slick presentation, a complete package, something that works, just like a CEO would expect in a company that took a chance on hiring you.

The problem with this is that the more time you give them, the less time there is to expose them to detail. There is a lack of depth. Sometimes the brush-strokes are so broad that students really don't get much content.

Since high school continues to build concept on concept, it gets hard to get to those challenging courses that gifted students crave. There will be a dearth of AP courses. At my children's school, they eventually offered calculus, but it was hard to get students ready for it. Too much foundation was missing.

Conversely, the strength of PBL is substantial. It is soaked in choice, challenge, and social connection—all the components of intrinsic motivation. Furthermore it requires a stand-and-deliver approach to results. You have to perform. It does prepare you for the real world.

My youngest is even more introverted than I am. Yet he learned to stand like a pro, look you in the eye, and tell you what's up. I love that!

Therefore, if you could lose the leeches, steer clear of the fluff, hold on to the motivation, and make the students mimic real life, you've got something. Here are some PBL basics.

The leech problem is an easy enough solution that I wonder why more people don't do it. John Dewey was a big advocate of what came to be known as *cooperative learning*. This system has individual accountability built in.

Here's an example that can be generalized to a method. If the project was to be about power distribution from generation plant to neighborhood, there are a lot of concepts to consider: *induction, safety, voltage, current*, and *electric power*.

A project could easily be set up involving models, a few calculations, some lights, etc. Maybe an adventurous group would make a circuit including a generator, a transformer and other sophisticated equipment. But no one should be hypnotized by the wow factor—especially the teacher.

One solution is to tell the group that at the conclusion of the unit each student will be responsible for scoring above a certain minimum on a quiz dealing with two or three central concepts. If this bar is not reached, the individual's project score can be adjusted down (or up, for excellence).

The second thing I like to do is keep things granular. This means, make projects small enough to fit into a day or two of instructional time. The advantages of PBL are still in place, but the tasks can be focused to keep learning advancing at a reasonable pace. One example is what I did today.

In my first year IB Physics class, they are learning about nuclear decay half-lives and nuclear decay equations. They also need some exposure to societal implications of nuclear manipulation. So, after exposing them to half-lives and equations, I assigned them topics to report on.

Usually they work in pairs or teams, but there were enough interesting concepts that I had each student choose from a list of possible foci, which included

- The history of Chernobyl
- The history of Three Mile Island
- The validity of the origins of the Hulk and Spider Man
- Carbon 14 dating
- Dating the dinosaurs
- Conditions for Nuclear Instability
- The history of the Manhattan Project
- The history of Hiroshima and Nagasaki
- Power generation in a fission reactor
- Power generation in a fusion reactor
- The work of Madame Curie
- How an A-bomb works
- How a hydrogen bomb works
- Wildcard

I actually wrote about twenty topics. I had students choose their top three to ensure they had something of interest left by the time they got to choose. Then I rolled the dice to randomize the selection order.

Wildcard added yet more choice. A few students took this option. One changed his self-selected topic from "the effects of radiation on animals" to "cosmic radiation exposure of passengers on long flights". I don't know where they come up with this stuff, but I love it that they do.

They had 40 minutes in class to research with our set of iPads. Any more time they felt they needed would be the night's homework. Presentations would follow the next day.

The only criteria I set for this assignment were 1) that they include a discussion of half-lives in their presentation and/or a communication of the nuclear equations involved and 2) that they entertain us. Note that the project centers on the standards *and* real performance. These two requirements will be connected to a rubric.

Sometimes, I observe students' visceral reactions to their peers' presentations. That's how those "oohs" and "ahs" of being entertained make their way into the rubric. I usually weight this less than *communication* or *demonstration of understanding*, but it still matters.

INQUIRY

How did you do science in school? Were all the labs focused on you coming to some conclusion the teacher expected you to find, and verified once you got there?

Inquiry goes beyond that. It goes hand in hand with how science is done. It starts with an open-ended question, ends with a communication of results, and may pick right back up again with a related question.

I am going to give the briefest of overviews, but with an emphasis on the fact that inquiry is a process that respects learner choice, either in how an investigation is conducted, what question is asked, or both.

I defer to Douglass Llewellyn. He lays it out in his book *Inquire Within, Implementing Inquiry-based Science Standards* (Llewellyn, 2001).

The inquiry cycle is comprised of six steps. See his book for details.
It boils down to getting your students to ask questions, then letting them craft investigations to explore answers.

Llewellyn presents a list of questions elementary school students generated while observing "ice hands" melting in a bowl of water. Here are some questions my students might ask while observing rubber balls bounce on a counter:

- Is compression distance related to bounce height?
- Why are some balls louder when they bounce?
- Does bouncing surface matter?
- How does drop height affect bounce height?
- Why do some objects not bounce?
- Are spherical balls better bouncers?

He also talks about teacher control. Weak inquiry happens when teachers make the questions. Strong inquiry has students asking.

Inquiry is great, but it is in constant tension with the volume of required standards. Doing inquiry right takes time, time teachers don't always want to commit to when a particular question rarely results in something on a state's list of required content—like *how transistors work*.

For inquiry to really take off someone big is going to have to convince us of the importance of the *skills* of doing science, putting them on par with the results we've assembled over centuries of applying that process. That's where the Next Generation Science Standards come in.

NEXT GENERATION SCIENCE STANDARDS

Maybe out of frustration with the question, "Why are our children lagging the rest of the developed world?!" we've made a move towards national standards. Now we have the common core of English and Math standards.

Science is not far behind. The *National Academy of Science*'s *National Research Council* rallied the states to form the *Next Generation Science Standards*. Here's the history, in acronyms: NAS→NRC→NGSS. Soon, science curricula across the land will be influenced, if not transformed by NGSS. So why not run ahead of the pack—especially if we like what we see?

So far, I like what I see. Much good educational theory is represented in the standards. Again, my intention is to expose you, the new teacher, to my favorite use of educational theory.

Still, it can get overwhelming, so I will continue to make connections among the various ideas, and I will summarize at the end of the chapter.

The standards come in *three dimensions*: *practices*, *crosscutting concepts*, and *disciplinary core ideas* (from http://www.nextgenscience.org).

Practices. In earlier standards these descriptors used to be called *skills*. These are the behaviors scientists use when they figure out how the world works. Another set of practices involves what engineers do when they design and build systems.

At long last! Now there is a clear distinction between the two ways we figure things out. One way *extends* our knowledge of the universe, the other way *applies* this knowledge to build useful things. Under NGSS, students will do both.

Crosscutting concepts. One of the few decorations I ever really put time into was a set of ideas I got from the California Science Framework. I thought these themes made a lot of sense and ought to be displayed in the classroom, making it easier for me to make connections among ideas. They were

- *Patterns*
- *Cause and effect*
- *Scale and proportion*

- *Systems and system models*
- *Energy and matter*
- *Structure and function*
- *Stability and change*

I went to the media center, got some construction paper, and pressed out these phrases. Then I searched out little graphic press icons that embellished these ideas. For scale and proportion, I used two different sizes of the same graphic. I found similar ways to display each concept.

Then I laminated everything and posted them so they ran along the crown of the classroom. I moved away from that school long ago, but I really should post the same phrases in the dungeon.

And I should do this not just because they are now the "Next Generation Science Standards' crosscutting concepts," but because they are so useful. The NRC understands how valuable it is that learners make connections among ideas, and I agree.

As I see it, this is one of the main functions of a teacher, to help these connections be made—and reinforced as often as possible. But I also know teachers themselves have varying degrees of 1) knowledge of the other science disciplines, and 2) practice making connections within and between disciplines.

Therefore, if we are going to help students make connections, we need practice making them ourselves. It would be a worthy staff development exercise to have science teachers sit in common disciplines and make explicit all the content (and practices) that connect through each cross-cutter.

For example, under *scale and proportion* a group of physics teachers might list the following:

- Volume vs. Length
- Gravity vs. Planetary radius
- Gravity vs. Mass
- Electrostatic force vs. separation
- Strong force vs. separation
- Nuclear stability vs. nuclear diameter
- Heat loss vs. hot object radius
- Wire resistance vs. wire radius or wire area

The next connection-making session could involve science teachers across disciplines. Working together, a physics, a chemistry, and a biology teacher might brainstorm under *energy and matter*

- A rock on a hill
- Enthalpy, and
- ATP

The same set could almost fit under *stability and change.* Under *systems and system models* these three teachers might also list

- A planet orbiting a star
- Bohr's model of the atom, and
- An ice skater in a spin (ok, that's a stretch)

So, if you find yourself in a position to recommend some collaboration, this is a good place to start.

Disciplinary Core Ideas. These are the content pieces the NRC thinks are important, have interest to students and society (Yay motivation!), and can spiral up in sophistication as they are taught again in higher grades. They are slightly different than what we are used to. For example, physics and chemistry are combined as *physical sciences.*

MISCONCEPTIONS

Sometimes you think students "get it" when they really don't. Yes, that's what quizzes and tests help you assess, but sometimes a kid's naiveté can find its way through undetected.

There is a science to exposing and challenging these misconceptions. James Minstrell has done much research in this area. It's worth having a look.

The trick to beating misconceptions is to expose them and create experiences where students can see they are wrong and know why they are wrong. Sometimes you have to be persistent.

One common misconception in physics is the belief that you need to keep a force applied in the direction of motion to maintain that motion. Of course, the truth is *the first law of motion*, that an object will keep going at the same exact speed forever unless a force is applied to it. No force, no change.

This one is tough, because students' life-long experiences say otherwise. One way to address the misconception is to use "frictionless" carts and have the carts push friction blocks placed in front of them. Students or the teacher can launch the carts with a spring-loaded plunger and record the time it takes for the cart to stop for each mass of blocks.

Next, they can graph "Stopping Time" vs. "Friction Block Mass" and see that reducing the mass causes the time to approach some really big number, like "forever".

The teacher could also show an animation of satellites in orbit, commenting on the fact that they keep going, though on a path curved by gravity.

My accomplished students' misunderstandings of entropy (see last chapter) might be considered a misconception too. Hopefully, the Anti-entropy Device Project sufficiently addressed that one.

So, in sum, repetition of a difficult concept in various contexts should bring about the right understanding. Sometimes, however, even basic foundational ideas need repetition, not to address misconceptions, but to obtain *automaticity*.

AUTOMATICITY

I had to throw automaticity onto my must-have list because it is a good counterbalance to time consuming methods like PBL and Inquiry—and because I believe in it.

Imagine a world where you have to look up your usernames and passwords every time you go on line, and your job involves jumping from online service to online application all day long. It could seriously impact your productivity.

That's what it's like when we ask students to read but they lack the vocabulary to keep up with the rest of the class. Automaticity demands that students can do important foundational operations quickly, automatically.

Of course automaticity is important in math and science too. For example, if I ask students to calculate the frequency of light given a certain wavelength, it's a simple division problem. But if they are unpracticed in scientific notation, dealing with those huge powers of ten is going to be a real struggle.

Often, students get shuttled through math without really understanding decimals, fractions, and proportion. These students run into a wall when they start doing *stoichiometry* in chemistry.

In physics, if I rush through the topic of deriving acceleration from speed, it is going to slow them down when they are learning the *second law* or when they are asked to find forces that arise from changing speeds.

Blasting through acceleration might also make students unclear about the operator Δ, delta, which appears again and again in physics, from *uncertainties* to *impulse*. In turn, applying impulse in the lab environment is going to be slowed-down too, compounding the difficulty and limiting the depth and quality of their experience in the course.

Getting automatic takes practice. You don't necessarily have to drill students, but they have to have enough problems in enough contexts that they recognize the situation they are in and can act. How to get them there is a matter of teaching experience.

I just encourage you to remain open to figuring out when the students have become automatic. Too much time, and you leave them bored and disengaged. A solution to boredom and disengagement is *differentiation*.

DIFFERENTIATION

Here's a scene I don't like seeing: Everyone's working hard. Then the whiz kid finishes fifteen minutes before everybody else, turns in his work—it's excellent—whips out his smart phone, and starts playing a mindless game.

My fault. I'm wasting his life. He might think he deserves a break for being so awesome, but he would likely appreciate a challenge even more. I know his future self would.

Differentiation is a blessing to gifted children. Where I teach, there is a Gifted and Talented Education program (GATE). I am told it was designed not so much to take our best and turn them into monster engineers so we could beat the Russians, but simply to keep the kids alive.

The stats were bad: really smart kids were killing themselves much more often than the rest of the pack, probably because they were so bored. GATE, incorporating strategies like differentiation, helped quite a bit.

I also like to accelerate students, entertaining the thought that this will help them achieve more in life, and maybe improve life for all of us.

Back to the classroom—the above situation is typical of a unit test. Here's how I kick the kids' butt with differentiation. I will offer extra credit. They don't have to do it. Anyone could still get a 100%, but usually, these kids will bite. Here's an example:

> EXTRA CREDIT/ALTERNATE PROBLEM
> 6. A heat pump takes 200J from a hot reservoir and dumps 120J into a cold reservoir while doing 60 Joules of work.
>
> a) Sketch a diagram for this
>
> b) Next to the diagram above, sketch a diagram for a refrigerator that violates Clausius' statement of the 2nd Law, and, when added to the first pump, shows that the result is a violation of the Kelvin-Planck statement of the 2nd Law.
>
> c) Next to the diagrams above, sketch the resultant diagram.

Part *a* anyone could do. Parts *b* and *c*, even I forget how to do!

Or it could be a problem much like the rest, just more complex. The first problem below was from the standard part of the test. The one after is the extra credit:

4. The diving board below has a uniform weight distribution. It has a mass of 40.0 kg. What is the force at each support?

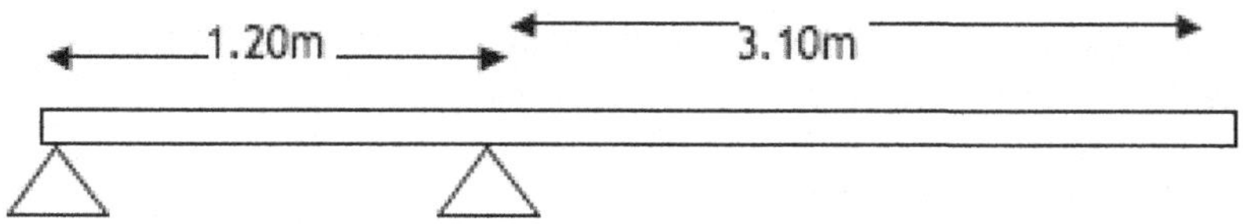

EXTRA CREDIT/ALTERNATE PROBLEM

Solve problem 4 with an 80.0kg diver standing at the end of the board. Do not neglect board mass.

One nice thing about differentiation is that it relies on important motivational components, *challenge* and *choice*, as mentioned in that chapter. For gifted students this often means that if you challenge them, they will choose to do it.

On classwork or homework assignments, you can always give the option for students to do a certain number of problems from the "challenge" section of the chapter. Most books have these.

Projects are easiest to differentiate. You might not have to specify anything additional to get the differentiation. They will surprise you with the complexity of their work.

Still, I sometimes build it in subtle ways. In the list of nuclear topics above, the whiz kid might bite on the most difficult topic shown like "power generation in a fusion reactor". Or, they might come up with something crazy under the "Wildcard" option.

In a recent list the hardest topic was "special relativity". There was a set of about three students in my advanced class who I knew would jockey for that one. And that's exactly how it happened.

One of my favorite differentiation strategies is to have students design some sort of device on paper. Being a *synthesis* question, this is an automatic differentiator, as well as a form of Higher Level Questioning.

HIGHER LEVEL QUESTIONING WITH BLOOM'S TAXONOMY

I was an administrator in only two of my 21 years in education. My better year in administration was when I functioned as Sage High's "Science Administrator". Let's not talk about my term as Vice Principal.

I was working with Liz, a colleague who was the school Literacy Administrator. Both of these positions were experimental, and only lasted a short time. The one year we were together, Liz and I helped design and implement that year's Professional Development.

This PD came out of our principal's vision for the year, *Higher Level Thinking*. (You'll also hear this called *Higher Order Thinking Questions*, among other labels.)

There are several reasons I appreciate what our principal implemented that year. First, our school ranking made a huge gain that year. Second, it used old educational theory, the tried and true *Bloom's Taxonomy*, aka, *Bloom's Major*

Categories in the Taxonomy of Educational Objectives (Bloom, 1956). Third, it was fun. Fourth, it was another differentiation tool. Finally, it allowed me to witness incredible things in classrooms.

Bloom's isn't just about questioning. It could be used to categorize everything a student did. It's helpful to remain conscious of the taxonomy—not to avoid any level, but to recognize when we *are* avoiding a level. Doing so gave us the opportunity to fix it.

Bloom's is all over the web, but I will summarize the categories and paraphrase the descriptors here (adapted from http://faculty.washington.edu/krumme/guides/bloom.html). Note that the highest level, *Synthesis*, used to be the second highest level, but since it is the creation category, most people now put it at the top.

Synthesis. Creatively applying knowledge or skills to produce something new—combines, compiles, composes, creates, designs, formulates, models.

Evaluation. Judging the value of material based on personal value or opinions, resulting in a product that escapes being labeled right or wrong—appraises, compares and contrasts, criticizes, defends, interprets, judges, justifies, supports.

Analysis. Breaking down information into component parts—trying to understand structures, developing divergent conclusions by showing causes or motives, finding evidence, making inferences, illustrating, differentiating.

Application. Using learned information in new contexts to solve problems having one right or one best answer—articulates, computes, determines, prepares, projects, solves.

Comprehension. Understanding the meaning of information—classifies, describes, estimates, generalizes, gives examples, makes sense, paraphrases.

Knowledge of facts, terminology, and conventions; remembering previously learned information.

Focusing on crafting questions from Bloom's was the way to go because it forced teachers to give students access at various levels up and down the taxonomy. We emphasized the higher levels because those are the most often overlooked. We figured that a balanced use of question types would drive learning deeper within students and wider across the student population.

Here are some questions from each discipline, with corresponding categories.

Physics

1. If a person could lift 220 lbs on Earth, what could that person lift on the moon? (Application)
2. How would the sport of weight-lifting be affected on the moon? (Analysis)
3. What are the units for gravitational field? (Knowledge)
4. Use what you know about centripetal acceleration to design a spaceship that simulates earth's gravitation. The period of rotation must be at least one minute. Make a sketch and diagram it with the most important information. Include bonus features that would open up more opportunities than those found on earth. (Synthesis)

Chemistry

1. Name an alkali metal (Knowledge)
2. Give a metal that reacts similarly to magnesium. (Comprehension)

3. A certain college wants to teach only the Bohr Model of the atom. Do you have a problem with that? Explain. (Evaluation)

Biology

1. Describe how energy flows through the community, as shown in your food web. (Application & Analysis)
2. What effect are cattle having on certain ecosystems? (Analysis)
3. Give an example of an adaptation. (Comprehension)

Among other things, we asked teachers to write test questions in all categories. We also planned lessons with teachers on how to give lessons incorporating prepared questions from all categories. We usually did this in three steps, 1) co-planning one lesson taught by the administrator, 2) co-planning another lesson, co-taught by teacher and administrator, and 3) co-planning a third lesson, taught by the teacher, with the administrator observing and giving feedback.

This planning and teaching cycle was most informative. I remember one lesson in particular. The teacher was demonstrating the making of waves on a string. He kept asking questions that gradually moved up to the higher levels. As he did so, students I had witnessed months and weeks earlier—those that tended to linger at the back of the room and goof off or just tune out—these students started migrating to the front of the room. They had no attitude, just blank looks of attentiveness. They started answering these questions and asking more.

When they were asked to analyze the wave, they attributed features to tension in the string, essential to speed and amplitude, but previously unmentioned by the teacher or anyone else. We pressed on to synthesis questions, asking about how other waves could be made. The engagement continued with these students.

This pattern of chronically disengaged students suddenly coming back to life repeated itself in various rooms. To me it was dramatic, the reverse process of the spreading of the zombie scourge! We were reclaiming them.

As administrators and teachers, we didn't talk enough about this one phenomenon. We made other tangential work for ourselves. But in retrospect I can't let it go. By nature, this style of questioning was pure differentiation, but it wasn't packaged up and fed to the gifted students like apple-spiced walnut cake. Instead, it was for everyone. Yet it worked to identify that dark-matter cloud of lost gifted children. So much potential!

MARZANO'S RESEARCH

I'm ending with Marzano because I am a maximizer, and so is he. Robert Marxano did a very smart thing. His book *Classroom Instruction That Works: Research-Based Strategies for Increasing Student Achievement* reported on the best strategies teachers use in their classrooms. He produced a list of nine that have the biggest impact on student learning (Marzano, 2001).

I'm not going to tell you all nine. You will likely run into one of his books at your school, or you will get a copy for yourself one day. OF COURSE some of the strategies are part of what I already shared with you. Another strategy, *feedback*, I'll talk about later.

Here I'll be explicit about two. I'll start with # 2, *Summarizing and Note Taking*. Have your students summarize. Have them take notes. I do this when I introduce a topic. His research shows that the more detailed a student can be in her notes, the better. However, if students are too conscientious about trying to get everything, they might miss everything. They can't process what they are recording. It gets totally counter-productive.

Drum roll please! Marzano's top strategy is . . .*identifying similarities and differences*. Venn Diagrams are great for this. Here is an example from my classroom.

Alpha, Beta, and Gamma Radiation Venn Diagram

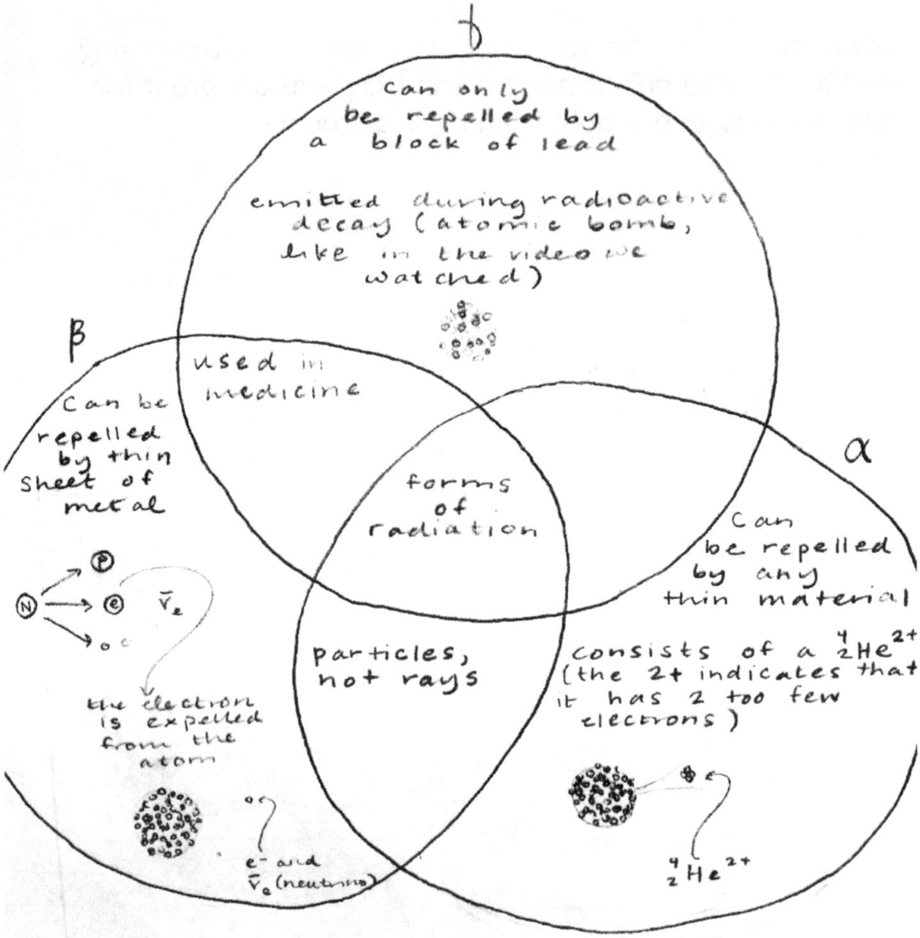

Summary:

Motivation, PBL, Inquiry, NGSS, Misconceptions, Automaticity, Differentiation, Higher Level Questioning, and Marzano's Strategies

Use these. Understand that *choice* is one of the most powerful common denominators across this set of ideas. Also, realize that freeing your students ***requires*** that you believe they can be trusted with the freedom. And they can. They will even take it to higher levels.

Finally, be patient with yourself. You can't do all of this in your first month of teaching. Do something simple on day one—maybe assign a Venn diagram!

6. Your Day

Years ago, I was walking stiffly to the bathroom on my prep period. It was peaceful, quiet. I ran into a social studies teacher who looked a little lost, like a schizophrenic who dropped his cigarette on the patio.

He said, "I hate these block periods. I get so bored on my prep." I wanted to squeeze his feet like Play-Doh through a press.

To be an effective science teacher you have to put in time. The pleasant surprise is that your upfront investment pays off in efficiency later on. My earlier self would never have believed my current schedule.

When I landed my dream job, teaching IB Physics at Sakura High, the guy I replaced confided that he had been spending sixteen-hour days preparing to teach IB Physics.

The problem was he was credentialed in biology. Physics was not his thing. But he was so conscientious that he wanted to deliver the content impeccably, every day. Wow. It turned out he had only been there for about a year, but a year of sixteen-hour days—I don't want that.

I was less extreme. I did push too hard in the beginning, and it started catching up to me. See the *Burnout* chapter. Also, ask my wife about what it's like to experience labor pains while her husband runs into Blockbuster to get the right science fiction video for the sub. I will show you how you can avoid that.

The biggest two time consumers are *lesson planning* and *grading*. I will detail how you beat those back to manageable levels in later chapters. You can use some methods for pumping up your efficiency immediately. Others can be phased in.

Here are two schedules. One I followed when I started teaching. The other is how I run my day now.

When I Was New

6:00-7:30	Before school	GRADE WORK/ PREP FOR LESSONS
7:30-8:30	Period 1	TEACH
8:30-9:30	Period 2	TEACH
9:30-9:45	Nutrition Break	EMAILS
9:45-10:45	Period 3 (Prep period!)	PLAN LESSONS FOR NEXT DAY, GRADE WORK
10:45-11:45	Period 4	TEACH
11:45-12:30	Lunch	SUPERVISE CLUBS/ EMAIL/ TUTOR
12:30-1:30	Period 5	TEACH
1:30-2:30	Period 6	TEACH
2:30-5:00	After School	GRADE WORK/ PREP FOR LESSONS & LABS

Present Day

7:00-7:30	Before school	GRADE WORK/ PREP FOR LESSONS
7:30-9:00	Period 1	TEACH
9:00-9:15	Nutrition Break	EMAILS
9:15-10:45	Period 2 (Prep period!)	PLAN LESSONS FOR NEXT DAY/ GRADE WORK
10:45-12:20	Period 3	TEACH
12:20-1:00	Lunch	SUPERVISE CLUBS/ EMAIL/ TUTOR
1:00-2:30	Period 4	TEACH
2:30-3:00	After School	GRADE WORK/ PREP FOR LESSONS & LABS

I've always worked through lunch, so let's say I used to work an eleven-hour day. This day has now become an eight-hour day. At times, I get home even earlier. The dungeon is kind of stinky, so I really get there later and leave earlier than the schedule shows. To make up for this I bring a little bit home with me.

Plenty of people work eleven or more hours in many professions. I don't mean to cry about it, but if you are not cut out for it, teaching saps you. You have to have the disposition to stay aware and to react quickly to events in your classroom. That eleven hours could be a long eleven.

Or a fast eleven. I like the stimulation. What they say about teaching, for me anyway, is partly true: "Those that do, do. Those who can't do, teach." I was a good engineer, got a patent, helped make the first optical fiber that could be used on spinal cancer surgery.

But I couldn't do it. I couldn't sit in a cubicle all day. And I couldn't commit my life to polymers. Introvert, yes, but that was too much. I liked all of science, and I liked explaining things. I thought teaching would work out for me, and it did.

What these schedules do not show is the rich diversity in my day. Every class period is different. My mood changes. The kids' moods vary. Maybe there are current events to talk about. Maybe something is really funny and we laugh about it.

If you are going to show up at six and leave at five, you'd better enjoy ghost towns, because that's what schools are like at those times of the day. I found them to be my favorite times of the day. I miss them.

One of the ways I've shortened my day is by establishing a routine. When I get to my prep period I go right to the lesson planning. That afternoon I'll open the box where I throw the turned-in work and I will grade as much as I can.

My current school has hour-and-a-half long prep periods. That extra half hour is golden. I will usually take a break in the middle of it, walk to my buddy's counseling office, and pour a cup of coffee.

When I was a new teacher, I really needed all eleven of those hours. In fact, I exercised discipline to keep them down to that. I brought nothing home. (Well, a little, sometimes.)

It takes time to plan lessons when you are new for many reasons: 1) You are not used to the process. 2) You might think you need to script every minute. 3) The curriculum might be new to you—see the biology teacher teaching physics, above. 4) You might need to scratch through the stockroom to see if you even have what you think you need for your demos, and 5) You do not have the advantage of having a previous year's lessons to work from.

Grading also takes time for similar reasons. More on all this in the *Planning* and *Grading* chapters. I'll also show you how to be more efficient from the start, so you won't have to do all eleven hours. Not that you want to be a slacker. You just don't want to burn out. You don't have to. Let's keep it fun.

Summary:

- **Quality science teaching takes time.**
- **Schedule a long workday when you are new.**
- **Be efficient.**
- **As you get more accomplished and accumulate ideas and resources, shorten your day.**

7. Control the Class/ Lead the Students/ Know Your Kids

Who is smarter, Louis C.K. or Stephen Hawking? That's a tough one, isn't it? While you might say C.K. is just a goofball—really? Who can think that divergently and bring it all together with so much irony that you *have* to laugh?

There has been much said about multiple intelligences, but the basic idea is that the newborn child has predilections that look like they come from a roll of the dice. They emphasize certain things. Maybe they crawl to the piano at a young age and plunk out a tune. Maybe they can throw and catch really well. I believe a lot of people are geniuses, but just geniuses with different criteria for intelligence.

A principal I used to work for took gifted education one step further. I mentioned the GATE program, but there is a smaller set of super-gifted kids who killed themselves at an even faster clip, and who kept on doing so *despite* the GATE program.

These represented not the top 2% in IQ, but the top 0.1%. He made a program for these kids. He called it *Seminar*, and it worked.

Now, if you thought the GATE kids had some quirks, when you brought the Seminar kids together it was clear. It was not uncommon to find kids in wheelchairs. Sometimes at birth a child is starved of oxygen and part of his brain dies.

Rarely, but sometimes, the part that dies is the part that controls motion, which is why they are in wheelchairs to begin with. But this frees them. It frees them to think. And when they think, their brain grows. Nerves get thicker, reinforced by the work they do thinking about things. Like a lifter squats twice his weight, these kids grow to think circles around us.

My point in mentioning these mega-geniuses is that sometimes it seems arbitrary how the brain focuses and strengthens itself. All kids are smart, but not all of them are smart in the math and reading way we traditionally define smarts. Therefore, we cannot be upset with them when they are not as quick as we would like them to be in science.

I mention this in the *Control the Class* chapter because I want you to understand that your students don't always jump through your hoops how you want them to simply because it is not as easy for them as you think it is.

They are not resisting you. In fact, they might be embarrassed, because they know they are not like others who are good at it. This embarrassment often leads to non-compliance because it is scary for them. It's like when my wife asks me to dance at a wedding. Horrifying.

It doesn't mean you never ask them questions. Just be patient and fill in some logical steps for them. But don't be mad.

You will notice that I give no technique for controlling behavior in this chapter. You'll find that in Chapter One. There, you will also find me referring you to a book on classroom management.

However, here in this chapter I am emphasizing the importance of you knowing your students. When you accept their differences, it will keep you from blowing things out of proportion and taking things personally. The psychologist Carl Rogers called this attitude "unconditional positive regard".

It took me years to fall into this mindset. I was delighted to find that there is a group here in San Diego that helps teachers improve their effectiveness by coaching them to achieve this unconditional positive regard. It's a group called Education Transformations (www.EdTrans.org).

Please look back at the LEAD THEM section of Chapter One. Here is the summary of that section and of Chapter Seven.

Summary:

- **Lead them**—Give them reason for what they are doing. Give them something they can work together with you to achieve. Tell them what you want. Address mistakes with fairness and consistency. Celebrate successes.
- **Accept their differences** from each other and from you.
- **Do not be threatened by their differences.** Do not think they are intentionally defying you by not obeying you.
- **But they actually might be defying you.**
- **Get that book on classroom management.**

8. Planning Your Lessons/ Planning the Year

Lesson planning lets you be creative each day. It starts with knowing what it is the students need to learn or review that day. Then it goes to letting your mind run to all the ways you can make that happen. Can you make it fun? Can you add some rigor? It ends with making choices that orchestrate the learning to accomplish the goals within your 60-minute period.

Yet before you can seize the day, you have to do the boring task of dividing up standards you want to address by days available. This comes from the very left-brained task of planning the year, usually in early September. If you do this annual work well, it leads you to the start of each lesson, as mentioned above, delivering the 'what it is' for each day going forward. It is a comforting thought knowing that you know what you have to accomplish each day even before you sit down to plot the course.

PLANNING THE YEAR, OLD SCHOOL

When you are new, it is hard to tell how much time to spend on any one topic. One solution is to let the textbook be your guide. Your district spent some time choosing the book. Likely one major selection criterion was how closely the book adhered to the state standards. So it's a safe bet that if you get through the book, you are going to cover the standards.

Our teacher education program matched us up with a veteran teacher who showed us how she did this. She took the number of teaching days in the year, looked at the number of pages in the book, and roughly matched instructional time with equal chunks of pages. Her next step was writing down chapter headings and topics in a yearly calendar.

Then she fell into machine-mode. She looked up at the book, slapped at some pages, then bent back down to jot micro-topics and page ranges in her calender—back and forth, back and forth until she got to the back cover. Done.

The sense in that is obvious—an inferred correlation between page numbers and amount of time needed to communicate a topic. As long as you factor in time for review and testing, that method can't be too bad.

The pacing might be a little fast or a little slow at times, but you get the job done that way. You also bypass the embarrassment of finishing the year only getting halfway through the standards. I have never known anyone to finish the standards early, so this veteran's solution kept the necessary pressure.

I could do the same thing, but I would first paper-clip the bonus material out of my search. Books often have too many chapters because they know states, districts, and teachers all have their favorite topics. You can't do all. Some are a waste of time.

PLANNING THE YEAR, HIGH TECH

Book publishers are desperate to win contracts, so they will add bells and whistles to their product. It's not too hard for them to include pacing guides. You can use these.

PLANNING THE YEAR, PERSONALIZED

This is my favorite method. I start with the state standards. In my state I get them at http://www.cde.ca.gov/, and they are just a couple of clicks away from there. I print the list.

Then I identify those few topics that I think are really central, and, frankly, that I like, and I make sure that I give myself plenty of time to do lots of good things in those areas. Then, I back-fill the days remaining so I still cover everything.

You might worry about how much you can skew the available time to your preferences. Don't worry. Start with the number of days in the year (minus testing and review). Divide by the number of all standards on the state list. Increase the number of days you allocate to your favorite standards. Don't go too crazy! Subtract those from the total. Take the remaining time and divide it by the number of remaining standards. Done.

At Gigantic High there was a biology teacher who loved whales (of course). He gave a lot of days to whales.

I have a love-hate relationship with electricity. Some years I choose to really lay it down. Other years I emphasize other things and know that I can just slap out circuits and static electricity like butter on bread and I'm done.

PLANNING THE LESSON

Step one: Look at your yearly plan.

Step two: Think how you can communicate the content and give students experience with it. Be creative. Will it be all lecture and student practice? Will there be a video, graphing, peer work, a lab, mini-lab? As you plan maybe take a walk to let your mind wander through the possibilities. Or, if this is not your first year teaching, look at how you taught last year's lesson. Did it work? Maybe do it that way again. I refused to keep a notebook of previous year's lessons for the first few years teaching a topic. This forced me to think of various ways to communicate the concept. Additionally, it allowed me to take advantage of the new resources that have been coming online. But do save every lesson electronically and in maybe in print version too. When you want to get more efficient, start accessing those records, so you can give the most powerful conglomeration of lessons for each concept you've taught.

Step three: Encode the day's focus as an objective and/or standard your administrator can identify when visiting your classroom.

Steps four through x: Gather the ingredients for a tasty soufflé.
Step x+1: Type everything up on a piece of paper, in a PowerPoint, a Prezi, or some electronic display board format.
Step x+2: Save it. Print it. Write it on your white board, or scratch it onto your chalk board, or prepare to display it on your smart board tomorrow.

Step x+3: Look at your plan. Anticipate any questions you will have or materials you may need. Consider what you will say as you lead students through the activities you've listed. Is there a demo you didn't think of when you wrote the plan?

Some people think they need to script the whole lesson. That's an excellent practice for your first week of school. If you are tempted to do anything beyond that, see the chapter on *Burnout*.

About step three. You might think displaying an objective is a stupid detail. After all, the student will experience what you have laid out for them, why do you have to steal the thunder? Because you have a boss. See the chapter on Administration.

If they come, administrators might stay in your room for only five minutes. The first two things they look for are 1) classroom control, 2) a plan. If you show them what you are doing, if they see it is a standard, and if it looks like you are on the way to getting there, that's about all they probably need to see.

Don't write the standard, and a five-minute visit might turn into a twenty-minute visit. You choose. Make it explicit or don't.

Here is a sample plan you might see in my classroom.

Lesson Plan, Physics, Period 1, Tuesday, March 3, 2015
Mr. Burtson, Rm 601, Dungeon

Standard 3e: ". . .entropy is a quantity that measures the order or disorder of a system. . ."

Objective: Students present anti-entropy devices and explain how energy added to the system decreases entropy inside and how it simultaneously increases the entropy of the universe.

Writing: List machines we use every day that are anti-entropy devices.

Announcements: Tutoring, academic league, field trip forms—First law of thermodynamics-and-entropy-quiz tomorrow.

HW Review: Sketches and paragraphs.

Prep: Take 6 minutes to prepare what you are going to say and who is going to say what.

Presentations: By roll of dice. Be ready. Submit report after presenting.

Homework: Page 230, #s 2-5; page 231, #s 11-13; study for quiz

It should be obvious what's happening. Here is an explanation of each component.

1. Lesson Plan
2. Standard
3. Objective
4. Writing
5. Announcements
6. HW Review
7. Prep
8. Presentations
9. Homework

1-3 These headers should be there every single day. They give guidance to the students, but more so to the administrator(s) who might visit your classroom. Chances are that they won't, but don't be caught without a plan. Not good.

4 This is a warm-up. It is a very traditional component, and like everything else in this chapter, your teacher education program already addressed this. But this book would not be complete without a thoughtful treatment.

I love writing. Research shows it is important. The only reason *writing* did not make the educational theory chapter is because those things are super important, and I didn't think this was quite on par with those elements. Hmm. Maybe I should rethink that.

It is said that "thinking is writing, and writing is thinking." We used to have a writing coordinator at our school. She was awesome. She got us into the habit of having students write at the start of each period. Can you imagine? From auto-shop to English, every period these kids thought deeply. Sigh. Yes I miss her. My tribute is this writing component, which I re-introduced this year. Students take it seriously. Sometimes we have fun with it.

In general, the warm-up is a regular feature because it gives the teacher time to take roll, and time for the students to settle from the transition. That does not mean it should be a device to distract the students. Sometimes it is the most instructional part of the lesson.

Other forms of warm up could be problems to solve, tests or quizzes to self-correct, graphics to copy and annotate, or Venn diagrams to make—whatever can be done quietly and individually, initiated by simple written directions.

5 Optional. If you have something important to remind them of, write it in the plan.

6 Optional. I give homework most days, but not all.

7-8-x. This is the guts, the creative core of things you line up for students to do. See the earlier chapters.

9 Optional. Again, most days. . .

PLANS FOR SUBSTITUTES FOR WHEN YOU ARE ABSENT

You are going to get sick, or you will be expected to go to a professional development. This will happen in your first year. Here is how you prepare for it: KEEP IT SIMPLE.

When I was new I was super-detailed in my personal plans for teaching. Must have been residue from when I wrote Standard Operation Procedures as an engineer for the biotech I used to work for. No No no no noooooo! to what I did with subs: I stupidly wrote the same detailed plans.

It was amazing that some left notes that they appreciated the detail. And I appreciated them. They did what I told them to do, so it was almost like I was there after all. Great!

Then there was the sub who just read the paper and did NOTHING. Or the better sub who put on a Disney movie.

I used to get mad at substitutes, called them "scrubstitutes." But I recently realized it is not their fault. They have been trained by the teachers, who do not expect the subs to get anything done. They count their absence as a lost day, so they treat the sub as a babysitter. And when a sub makes only $100 a day, that's almost what a babysitter gets! Ask my daughter.

So, I try not to be absent. It helps me anyway, because when I retire, those accumulated non-taken sick days get credited back to me. Some people retire a little early on these!

But other teachers use every sick day available to them. If not truly sick, they call these, "mental health days," and I fully understand why they need these and why they are indeed medical, or medical-y. But I try to come to school every day.

Here is how to write a substitute plan:

> **Lesson Plan**, Physics, Period 1, Tuesday, March 3, 2015
> Mr. Burtson, Rm 601, Dungeon
>
> **Give worksheet.** Tell students you will collect it for a grade by the end of the period.
>
> **Collect worksheet.** Leave it for me to grade.

This is a golden plan. If nothing else, it gets students doing something they need the practice on. You can even put differentiated problems at the bottom! Call them "bonus" or "challenge" or "extra credit".

Once I left that plan for the substitute, but he pulled up a movie from his Netflix account anyway. Oh well, don't take it too hard. You can compensate.

EMERGENCY PLAN

Sometimes you will be absent without planning to be. Think. Think of a worksheet-style assignment for each of the subjects you teach. You do not know when during the school year you will be spontaneously absent, so think of something that would be a good exercise for them WHENEVER you are absent. Type this up sub-style. (See the section immediately above.) Put the plans, WITH ENOUGH WORKSHEETS FOR EACH STUDENT IN EACH PERIOD, in a folder you keep in the main drawer in your desk. When you leave school each night, straighten your desk, clear a large area in the middle of your desk, and plop down the folder. Make sure it is labeled "Lesson plans for today". Breathe easy.

Summary:

- **Plan the year mechanically.**
- **Plan the lesson creatively.**
- **Plan for subs SIMPLY.**
- **Plan for emergencies with general assignments that can work whenever.**

9. Administration

TENNIS BALLS

All the teachers crowded into a crescent shaped ring. The principal was alone in the center and started talking into his microphone. Just as he turned to look across, a couple of young teachers in the back hurled tennis balls at him!

This was early in my career. I was still trying to figure out what my relationship to principals should be.

Somehow no one saw it but me—not even the principal. The balls missed and bounced off. He kept right on talking. I was shocked. Were these guys buddies with the principal, just messing with him? Were they horribly insubordinate, not caring if they got fired? Is the school out of control? Are teachers upset about something? These guys were behaving like the worst behaved students in class!

It turns out they were just behaving like the worst kids in class. There were no issues. But the principal did not command respect.

WHAT IS A PRINCIPAL?

Teachers don't get promoted. It doesn't matter what you teach, how well you teach it, or what grade you teach. The only things that determine how much you get paid are your years teaching and your level of education. Then you top out. There are no bonuses. There's nothing at all wrong with that, as long as you know ahead of time what you are getting into.

Administration is different. You get more for being a vice principal. More still for being a principal, more still for being at a bigger school—maybe more for being a central office administrator, and of course, most for being the superintendent. On the downside, you can easily get fired.

So principals are on that path that rewards performance and punishes failure. As a result, principals can be some of the most interesting adults on campus. They will take chances. They will cuss and swear and get upset. Or they will stay cool no matter what. The better ones will wrap their heads around what it takes to really improve a school, and they will commit to that. The worst ones will just be paranoid all the time that they are being judged by the teachers below them and the administrators above them—and the parents. These principals get really reactive and weird and dangerous.

The principal is your boss. The principal leads the school. But some teachers believe principals are the enemy. Others want to be the principal. There is a spectrum. And, there is a way to treat principals that will be in your best interest.

REBEL, TEAM PLAYER, OR SUCK UP?

In my pre-teaching profession, I went to a seminar called "Engineer as Manager." I never became manager. But I learned a lot.

They taught me some simple basic truths that now seem as plain as remembering to breathe. In business, your boss has aaaaall the power. He can fire you. She can hire you. She can give you a raise. He can be disappointed and ask you to work till 9:00 every night until you solve the problem.

And your boss is boss either because he owns the company or because the board of directors gives her all the power. So, do you want to resist?

Therefore, if you get ten emails, which one are you going to answer immediately? YOUR BOSS'S. If you have three projects in your in-box, which one are you going to hit first? YOUR BOSS'S.

But you might not feel like being so accommodating at a school site. As a teacher you have tenure. They can't touch you, because once you have tenure, you really can't be fired, or demoted, or penalized, or asked to work late. You have a lot of power! Pretty soon you start seeing your boss as a schmuck. You have no respect for him, and you hurl some tennis balls at him on the opening day roundup.

It's a disgrace. It's not a disgrace when you truly have a schmuck. Who wants to give time to someone who can't think productively, act, or have integrity? But it is wrong to assume that this is who you will always have for leadership. It is also wrong to treat a new administrator as if he could never learn how to lead. It is hard to lead when you are a new administrator, just as it is hard to teach when you are a new teacher.

Instead, I like to leave the benefit of the doubt on the positive side. What if this principal really knows what the students need to get them excited about being in school? What if what the students start doing gets the attention of colleges, or teaches students what it means to start their own endeavor, or take ownership of a major project? What if this principal can convince the staff of her sincerity, and people start to follow with their creativity tapped and flowing? Such a person deserves respect, attention, and a chance to accelerate into full implementation mode.

Conversely, if everyone at the school slams on the brakes by default, the only hope for positive movement is eking out something good in one's own classroom.

Indeed the classroom is where it starts. And no matter where you end up or whatever craziness is happening outside your four walls, your classroom is your primary focus. A well-run classroom is the foundation in Maslow's hierarchy of teacher needs. But there could be so much more. *That is why I default to support for administrators, rather than opposition. And I recommend that for you.*

Once, a long time ago, that was my stance. There was a new principal. He was fiery. He had good ideas. He changed things. He secured lots of money from who knows where.

He continued to do wild things and occasional stupid things. In sum, I think he was definitely in the plus column, so I continued to back him up. A friend called me a *sycophant*. I had to look that word up.

Eventually this principal got fired. However before that happened, he turned on me, mocking me in public. But I still do not regret that I held out hope for the principal, even if I did so for too long.

TWELVE PRINCIPALS

I have worked for twelve. At this rate, by the time I retire, I will have had eighteen. Some of them left no outstanding memories, good or bad. A few were awesome. Some made major mistakes. It's probably much like management anywhere else.

There are stories from ancient times of kings and how you interact with them. In one account, if you walk uninvited into the king's throne-room, you are put to death. Your only chance of survival is if the king decides to extend his scepter. If he does, you may enter and live. He still might be ticked off though.

Principals are busy. If you have an opinion and you want to bug the principal about it, go right ahead, but I like to stay out of their way. I do not try to make him my friend. I do not try to be her enemy. I do my job, and if the principal wants something, he will let me know.

Regardless of your stance, your principal will get to know you. She will learn about you from talking with other teachers, from what students and parents say about you, and from observing your actions in the classroom and at staff meetings. If principals notice you in a positive way, they can support you. They can give you opportunities. If they notice you in a bad way, it is not true that they are powerless over you. They can make your day a little less pleasant. Or a lot less pleasant.

I have been fortunate. Each of my twelve principals have done nice things for me even though things have not been smooth with all of them all the time. Here is a brief history, starting with principal number 1, in chronological order.

1) Read my grant application out loud at a staff meeting.
2) Used to tell me his favorite demos (ex-chemistry teacher).
3) Met with me and encouraged me in my own dabblings as an administrator (later in my career).
4) Sought my opinion on things. Trusted me with some of his ideas.
5) Let me be a resource teacher, even though it meant she would lose a science teacher.
6) Came to trust me completely when I was a part of her administration team. Respected my ideas. Let me work with interesting colleagues.
7) Gave me free reign when I was his accreditation coordinator (WASC). Gave me someone very competent to help organize the process and the report.
8) Mentored me when I was attempting to get back into administration.

9) Hired me even though it meant rescheduling my interview and calling me when I was in Europe.
10) Trusted me completely in work I did as one of his vice-principals.
11) Worked with me on the band fundraiser concert.
12) Also worked with me on the band fundraiser concert.

On top of that, ten of these twelve hired me directly. They chose me. And that will likely be your case too. Sure you might work somewhere when a principal retires and is replaced, but chances are that most of the time the principal will interview and select you. It is important to keep that in mind.

WORKING FOR AN EFFECTIVE PRINCIPAL

If you are new and jobs are scarce, just be happy to start working. Get a job. If however, you have choice, it will benefit you to work with the right principal. Do some research. Do you know teachers already working? You will find that a teacher likely knows the reputation of principals of several schools in your district.

If you talk with a few of these teachers, that might give you enough principal-info to figure out which interview would be your most important one. Do your best to get hired at that school.

I eventually figured out how important it is to work under a solid principal. Not too many years ago our district did a major reconfiguration of three of its large low-performing urban high schools. They got tens of millions of dollars to break these three into fourteen small schools.

This move generated and required a lot of energy. Oprah Winfrey and Bill and Melinda Gates visited the campuses. Consultants and administrators came from across the country. All teachers participated in the redesign and re-applied to teach at the school of their choice. A net need for eleven new principal positions opened up.

I applied to be principal at one school. I didn't get hired. Instead, I took a science administrator position at a traditional high school. It was frustrating, because I believed deeply in this small school movement. Now I wouldn't even be able to teach in one.

The foundation of the movement rested on three pillars: *rigor*, *relevance*, and *relationships*. I finally realize in retrospect that these are basically the three factors of motivation, outlined in chapter four: *challenge*, *choice*, and *social connection*. No wonder I was attracted to the movement.

Luckily for me the principal of the high school I landed at was not going to be left out. Maggie Mason diligently worked the *three R's* into her school plan. She even reconfigured her entire Ninth Grade into small, personalized teams. It was an exciting year. That's where we did the Bloom's coaching.

Maggie pushed the Sage High staff hard, and she raised some controversy, but it was stimulating. More importantly, it worked. Test scores and school ranking shot up.

At the end of the year funding for my position dried up. Maggie kept me on as a teacher, and still included me in planning staff development, but I had my sights on getting back to small schools. I enjoyed being a part of something that worked at Sage High, and I wanted more of that, especially in the context the three R's work best, small schools.

This also became the first time I actively pursued teaching for a particular principal. This was a good strategy, because this next principal became the example for this section on *working for an effective principal.*

Steve brought me to DARYL High to teach math and forensic science and to lead the accreditation process. This gave me time to dig into the data that would show how well DARYL High was doing.

Since data on its own is meaningless, I compared DARYL High' success to each of the fourteen small schools. I also compared it to the success of the remaining comprehensive high schools in the district. Finally, I compared it to the average progress of all high schools in California.

I didn't have to dig that deeply. DARYL High was blasting off, by far the biggest gainer of the six in the original Sakura High complex. It was one of the top two of the fourteen district wide. It leapt several deciles, blowing away the comprehensive schools' and state's gains.

Curious, I tried to figure out why. A local university was studying the movement. Each year they surveyed all the students at each of the high schools. I remember that there were over a hundred questions. There had to be, because they designed the questions to gauge the degree the schools were implementing *rigor*, *relevance*, and *relationships*.

Two leaders appeared in the study. One was a Digital Media school on another campus. Its students ranked it highest of the fourteen on how the school successfully infused *rigor* into the curriculum. The other was DARYL High, ranked highest on *relationships* (also called *personalization*).

Surprise! These two were also the highest on performance gains. Maybe there was a cause and effect relationship there, but I attribute the gains to a more fundamental variable: leadership. Here is how Steve did it. If I were looking for a place to teach, this is what would attract me.

Vision. DARYL High's theme was leadership. Its mission was to improve student success. It valued fun. Whether or not I'm attracted to these specific foci, I can get behind a worthy vision, and this was that. Steve tracked DARYL High's student sports involvement as a percentage of the all six school-complex population. He did the same for involvement in school government (ASB). He had targets.

He also had goals for yearly testing: individual courses, overall scores, and school ranking. Steve kept each of these elements in front of us at weekly staff meetings. He had students make posters to line the halls with leadership quotes. I always wanted to put up a quote from Zack de la Rocha: "When ignorance reigns, life is lost," but I never got around to it.

The students, the staff, even the principals of the other schools all knew DARYL High was about leadership and improvement and fun.

Accountability and Feedback. Whereas the typical high school's weekly meetings are more like obligatory in-the-flesh announcement sessions, DARYL High's meetings were exciting. You knew you were either going to get some acknowledgement or be on the hook for something you were supposed to do, or both. It was a time where input was welcome. We had a joking, family atmosphere. Steve was a part of it, not off at a distance. He was more likely to throw a tennis ball at you, than you at him.

Risk. Steve promised to do certain things if we met specific targets. He played in the student/staff flag football game. I still remember the principal and athletic director connecting on a long pass to the end zone, sealing a lopsided victory for the staff. Most importantly, he lacked the capacity to hide his emotions. If numbers came in DARYL High's favor, he rejoiced with us.

Loyalty. Steve trusted his people. He sought good people in every area, and he let them have responsibility. In turn, he supported us. I saw him go out of his way to help a teacher who wanted to train himself in another profession, even though it meant he would be leaving teaching and the school.

In summary, this principal gave us a reason to come to work, let us know how to contribute, was fully in it with us, and stuck by us no matter what. Yeah, I'd work for that guy. And so should you.

WORKING FOR AN INEFFECTIVE PRINCIPAL

In the case of Steve, I was able to describe a principal who was running on all cylinders. None of my other eleven were missing on all of them. And I'm still getting to know my newest. There is no anti-Steve. The Sage High principal was interesting and exciting in her own way. She had some enemies, and propelled her vision differently, but overall, it worked well.

I will briefly describe warning flags and move on. I simply describe the inverse character.

No vision. There is a new principal in our district. He has every front office person answer the phone saying, "Hello, this is ABC High, the greatest high school in the universe!" I love that audacious talk. Let's see what he has to back it up.

Clear vision is rare at school sites. When it's there, great. When it's not, make your vision a personal one, like getting the best scores, or having the most number of science fair entries ever, or laughing with your kids daily.

Unclear Accountability and Feedback. At a minimum, principals will visit your classroom a couple times on the year you are being evaluated: Each of your first two years until you get tenure, and every other year or so after that. The better ones leave you a script of what they heard. The best ones include comments. The worst tell you nothing. I actually don't mind if they don't visit. It shows they trust you.

Muted Risk. Just CYA.

Questionable Loyalty. You do not feel like they will work on your behalf, or even that they notice you.

Craziness. You don't know why, but they act vindictively and splinter the staff.

So, what can you put up with? If they have all five of these failings, you might want to file for a transfer. Otherwise, there are other reasons to commit to a school site: the students, the staff, the curriculum, etc.

WORKING FOR A PRINCIPAL WHO DOESN'T THINK YOU ARE EFFECTIVE

Yep, it happened to me once. After my year at DARYL High, a very well known mover and shaker in a different district was asked to open a new high school, Gigantic High. This principal pursued reputable teachers. She invited me to apply.

While I felt bad about leaving something good at DARYL High, here was an opportunity to go to another exciting challenge, with possibly even more resources aimed at making things unique and special and awesome.

And things *were* different at this school. The principal also liked Bloom's taxonomy, working it methodically into lessons throughout the year.

But I did not resonate with two new features: 1) that each teacher was to give the same tests on the same day as every other teacher in the same discipline, and 2) a hierarchical reporting structure from the department level up to the principal.

The principal gave us certain directives. I appreciated the challenges: crafting Bloom's assignments and tests. But things did not go well when we met as a department. Whereas I usually have great relationships with peers, there was always tension in the science department at Gigantic. When we were supposed to *zig*, word got back to my principal that I *zagged*.

Repeated messages communicated that I was not welcome. I did what I recommend you do: Leave. I interviewed at *Buckler High*, just across town, and I loved it there.

At Buckler's first assembly, it was clear the students were in charge of the show, and it was remarkable. They even had a student band covering the Ramones' *Blitzkrieg Bop*.

In retrospect, I am so happy that Gigantic was not the first school I taught at. Had it been, I would have felt a total failure. I would have quit teaching. I would never have seen my kids achieve at high levels, and would never have had the chance to do the MMA Box Dive, Smash Lab, or Roach Launch.

So if that is your experience, go somewhere else and don't give up on yourself. If it happens again and again at your next two schools, maybe consider that you are not one who can teach, but one who CAN DO other things.

HOW TO INTERACT WITH PRINCIPALS

Be an asset to your kids. Be an asset to the school. Figure out the principal's vision. Support it. If it is not clear, do awesome stuff in your classroom regardless. If you are called to participate in anything the principal requests of you, be there with bells on. But speak the truth humbly. Don't be a sycophant.

Summary:

- **Respect your leaders.**
- **Know what to look for in a principal.**
- **Seek employment with effective principals.**
- **Support the vision, starting with your vision for your classroom.**
- **If you are not appreciated, look inside, learn from it, and leave to work elsewhere.**

10. Persona

When I was eighteen I was still trying to find myself. I was so emotionally naive that I didn't even *know* I was trying. That's why I think it's silly when people say going to college gives you a fresh start on being someone different. You gotta know who you are before you can be someone else!

I wouldn't advise trying to be someone else anyway. Happiness is a goal for me, and I think being true to myself is a cornerstone of happiness.

However, in teaching it helps to be bigger than life. So, how do you do that without faking your personality? YOU MAGNIFY IT. Don't find a new station, just turn up the volume!

In chapter one I mentioned my introverted preference. And it's true. When I turn down the volume, I'm in front of a computer screen messing with Excel all day. Happy.

My awakening into persona happened when Savannah told me to show my excitement about the lice combs demo. That's when I started turning up the volume.

But the volume had to go up on the side of myself that had an outward focus. That was my desire to show people amazing things. It is what compels me to teach. Turn up the volume on my introvert side, and I'm writing poetry on my computer screen.

So, I find myself stomping around the classroom, making objects into extemporaneous demonstrations of physical phenomena. Staplers become vipers. Meter sticks become levers. I would characterize my persona as *The Manic Enthralled One*. It's just an aggressive version of *The Nutty Professor*, but I don't like lab coats, moustaches or teased out hair.

I gave plenty of examples of how this manifests in my classroom (see chapter one and chapter three). Here are some other teacher personas I've seen. Most of these are science teachers, but not all of them: The Nutty Professor, The Crazy Aunt, The Cantankerous Uncle, The Helpful Big Sister, The Bad-ass, the Hard-ass, The Happy Director, and the Empath. There are many possibilities.

By now you know yourself well enough to turn up the volume. So get cranking.

Summary: **Amplify yourself.**

11. Feedback

I learned to speak better at a Toastmasters club. One indispensible part of their meeting is the evaluation time. It comes immediately after that week's speakers. The evaluator's job is to give the speaker a subjective evaluation, but more importantly, to focus on those particular areas the speech targeted that week. It could be gestures, intonation, or having clear points.

This method is effective because the speech work has focus. Therefore the feedback can address that focus. However, the immediacy of the feedback is the biggest advantage. Speaking is scary. Adrenaline flows. Recent brain research shows that adrenaline cements in memory the events happening during stress. Therefore if Toastmasters gives feedback when people are still shaky with adrenaline, it boosts the positive effects of that feedback, driving learning deeper.

Educational research backs that up. *The quality and the immediacy of the feedback are important to learning.* It is rare you can grade fast enough to get results to the students while they are still jittery from the test. But if you can get it back the next day, that's way better than a week later.

But that's a pain because warehousing stacks of papers to be graded is one of the ways teachers smooth out their day. There might be meetings and lessons needing planning on one day and not so much the next. That's when the stacks come out for grading.

I don't want to give up the warehousing option, so I rank the importance of immediacy of feedback here. Some things can go into stacks. In the next chapter, I'll break down other aspects of grading.

TESTS

These are high-stakes activities. Students know they affect the course grade big-time. Since I carry concepts forward, I make them aware of the learning opportunities that come from reviewing the test. They know this anyway. But I try to keep their focus less on their score and more on why they missed what they did.

For this reason, I will take the time to get the test done by the next day. If someone really wants to know, I tell them to come back at lunch. If I've had no prep time in between, I'll grade it right then. It doesn't matter. It's work I'm going to have to do sooner or later.

Students appreciate this. This year I heard one of them say, "Yeah, I know Burtson's going to have this done right away." If I establish this pattern, they start looking for that finished test, and they learn from it.

Being done right away makes it easy for me anyway. I remember which problems they had the most trouble with, so I learn too. It also gives me the points I need to go over the next day. So, my review adds to their own processing.

QUIZZES

Students still really care about quiz grades. If it's a short quiz, I can grade it right there on the spot, while the students are finishing up. Talk about instant feedback. Otherwise, the next day works for these too.

Some people like students to grade peers' quizzes. While this is a big time saver for me, I don't want to take up class time doling them out and re-collecting. Plus, it adds another layer of cheating you have to guard against. So I don't do this. However, if you like this method, it gives instant feedback. Maybe that's more important than the time lost and the occasional cheat.

PROJECTS

Use a hybrid approach to grading some projects. Here's what I mean. I wrote above about the ice cube challenge. Students want to know who won. I want them to know why.

So, I will have them show me their results that day. I make the report due the next day. When class time comes I gather the reports. Then I show the drawings of the top two finishing groups and compare them to the lowest two. Usually the top groups in this challenge figure out the best use of the paper is to wad it up and put it in the middle beaker so that it stops convection.

Students understand conduction, and some of them get radiation, but convection is the one that benefits most from this review. Showing team ranks matched with sketches as evidence makes it clear that convection is a real phenomenon. Of course, this kind of quick feedback works with other projects too: egg drop, parachute drop. Once you are done with this quick review, you can toss the reports in the file to get around to them later.

LABS

I still like to get these back quickly, but I will let them linger in the stacks if I need the time for something else.

HOMEWORK

I review it daily, so feedback is instant anyway. I have a simple homework grading method. See the next chapter.

CLASSWORK

This can go to the stacks immediately.

Summary:

- **Get tests back the next class meeting** and review the most frequently missed problems with the class.
- **Do the same with quizzes**, but do them that day, if possible.
- **If projects have a competitive component, show the class the champions, the last place finishers, and the design features that sealed their fate.** The rest of the grading can wait.
- **Labs benefit from quick feedback**, but can wait.
- **Review homework daily.**
- **Class work can wait.**
- **If you like being perfect, get everything back by the next time you see your students.**

12. Grading

Why do we grade stuff? My daughter goes to the University of California at Santa Cruz. Years ago, they decided NOT to grade things, but to give a verbal evaluation instead. Poppycock. They went back to grades.

If you are not holding your end of the bargain, you need to know. How else can you get better? So, I think we owe it to students to let them know if they are an A or an F student. So I do. But we have to be reasonable.

BEING TOO HARSH

Some teachers have fixed percentages in their minds: "I MUST fail 30%, or my standards are too low." Really?! We had a teacher at our school last year. She failed 50% of our kids. But our kids are super-smart. They come here for the advanced curriculum. But she had that number in her head. So now they have to make that up. Kind of ridiculous.

BEING TOO EASY

If anything, that's my problem. If you are too easy, there is no challenge. There is nothing special about your *A* if anyone can get it. And if anyone can get it, that de-motivates you.

You might think there is a science to giving grades. There's not. If there is, it's not just. Some people are still using the bell curve model: 10% A and F, 20% B and D, 40% C, no matter what the class or what the context.

People have tried to connect grades to assessments of standards learned. If you want to try that, give it a go. After a few years of trial and error you will find a system that works for you.

I admit it's hard to set up a system that is fair in how it assigns an *A*, an *F* and everything between. It does take experience. One of my biggest guides is *effort*. If a student is trying, he deserves some credit. And yet they might be trying as hard as Hercules. They might be pushing that rock like Sisyphus, but I wouldn't give them an A for it unless they exceeded the norm. But with that kind of effort they should be learning something—likely enough to at least *pass*. Effort matters. Effort itself is a sign of potential.

HOW MUCH TO GRADE

Some teachers see grading as their job, like mail to the carrier, milk to the dairy farmer, and widgets to the industrialist, they see items graded as justification for their pay. Do not fall into that trap.

These are the reasons to grade work:

1. To give feedback so students can improve their understanding.
2. To assess the degree of understanding so you can modify instruction.
3. To indicate how much a student has learned.

How much does it take to do that? You don't want two weeks to go by without grading anything. You also don't want to grade something every day.

I have one student. He is really, really small. He doesn't speak English well. He is a freshman. He is quieter than dew condensing on a willow leaf. He sits in the back, where people just don't do well, either by cause or by choice. His grades reflect that.

HALF WAY into the year, he started doing better. The last four tests and quizzes he scored like the geniuses. Now I'm scrutinizing everything he does. I'm thinking, "He's the real deal. We gotta get this kid to the university." I can't wait until I give the next test, so I can see what he does! I am super excited for him even though I think he is still oblivious to the significance of his turn-around. Maybe not. Something's working for him.

This excitement would not exist unless I graded often enough to see the pattern. But don't grade too much! Please pay attention to my second-to-the-last chapter on burn-out. Life is too short to spend it on evaluation.

Over the course of a semester, I would say the correct number of graded items is between 15 and 25. Too few, too many, and there are problems. There are roughly 18 weeks in a semester, so that puts you at just over one seriously graded assignment per week in each of your classes. The credit given to homeworks and verbal quizzes—I don't even count those, that's easy, though they do figure in to grades. Think about college. My grades consisted of far fewer than 15-25 components. How about yours?

HOMEWORKS

I assign homework almost every night. I credit them the next day by walking through the classroom as students do the writing warm up. I have them show me their work. I dedicate about two seconds per student to scan for degree of completion. On the seating chart, I will write a 1 (the number one) on the acetate with a vis-à-vis marker if they DID NOT DO their homework. I will write a "1/2" if it was half done. I will write nothing if complete.

After ten or twenty homeworks have been assessed in this way, I will enter the grade as a number out of ten or twenty in the grade book. But I give them one freebie. So if 20 homeworks have passed, and there are two marks on their name, I will drop one, meaning they have done 19. The program will award them 19/20 points, or 95%.

VERBAL QUIZZES

Sometimes I'll say, "I'm not checking homework. Instead, we're having a verbal quiz!" I will roll the dice to choose a responder, and I will ask for an answer from the homework set. Or, I will ask how they solved a certain problem. I will keep this up for five or more responders. If they get the answer easily, I will write a 10 on the seating chart. If they need more and more help, that grade will drop to a 9, 8, 5, etc.

Verbal Quizzes do not have to be relegated to only replacing homework checks. I usually announce when there will be a general verbal quiz the next time we meet. I tell students the topic in advance, so they will study. Sometimes verbal quizzes are unannounced.

One of my favorite times to give a verbal quiz is when I am trying to get them to understand something importantreespecially when that thing really can't be learned easily by just reading about it.

For example, the right hand rules require complicated descriptions of charged particles flying through three-dimensional space. The rules give the direction of the force on the charge. This direction also depends on the direction of the magnetic field.

Working in pairs, the students practice positioning their hands. If you were to watch the class without knowing what was going on, it would look like I had them standing up doing Tai-Chi. It's really confusing to them where they should orient their fingers, their thumbs, and their hands. Then, one by one they start to get it.

Very soon after that we'll do the verbal quiz: "Jackie, the field is pointing up through the floor and the proton is flying from the front of the room to the back, what is the direction of the force?"

There is a pause. She positions her hand, "To the left."
"Correct, ten points." The feedback is instant.

In a second problem I ask, "Joe. . . what is the direction of the force?"

Up comes the hand, "Down".
"Think again, it's an electron."
"Oh, up!"
"Good, nine points."

I don't verbally quiz to motivate (extrinsically) everything that happens in class. But again, for something like this, there is no room for, "I'll practice this at home." I know the trouble they'll run into, so I create the urgency with the verbal quiz.

QUIZZES

I keep these as little formative assessments, just checking understanding with a concept or two. It's easy enough to make two versions of the quiz on half sheets of paper. When paper gets scarce, and if the problems lend themselves to it, I will set the quiz under the document camera.
This week I was able to fit both versions side by side. I labeled one "girls" and the other "boys". I had them change seats so they were not sitting next to someone of the same gender.

Last year I caught a boy trying to pass himself off as a girl on a quiz—easy to catch the cheating that way! I also used the same quiz in a different period, but switched which set was girls'—all to shut down cheating.

I make it add up to 10 points. An 8 is an 80%, etc.

TESTS

I like to give three to five of these a semester, counting the final. I assign points to each problem, making everything add to 100. The only thing that might differentiate my tests from some other teachers' is the differentiation, as mentioned earlier. That includes having challenging questions as extra credit options.

After all these years I still get some test averages that are too low or too high. However, I know what they are supposed to know by a certain time. So if the average is too high, I am really happy with that. I won't curve down. If it is too low, I blame myself, and try to review later. I will curve up if that's the case. (Hey I had some engineering exams in college with a class average in the 20's! I was glad for the curve then.)

As I said above, I don't believe in the standard bell curve. There are too many variables. I was not able to find a program on line that curved how I liked, so I made one in Excel. If you want a free copy of my "curver", email me at burtson@cox.net.

Any of the topics in this book can be expanded on. There are lots of connecting ideas. One of these peripheral testing points that I must mention is bi-modality. If you have a cluster of kids around a high test average and a cluster of kids around a low average, that's a clear message that a certain percentage are not getting it. Unless you address this, that low bulge will get bigger through the year. No worries. Just keep an eye on it.

RUBRICS FOR LABS AND PROJECTS

If a piece of work has a lot of moving parts, you have to find a way to give a fair and informative mark to it. That's what rubrics do.

You have likely had some practice making rubrics in your teacher education program. I'll share a couple of thoughts to give another perspective.

Labs. I really like the way IB does their lab grading. They have us assign a *zero*, *one* or *two* to particular aspects of a larger criterion. These numbers stand for "no evidence for," "partially complete," or "complete". (These are not exact quotes.) For example, *gathering sufficient data* might be an aspect of the *data collection and processing* criterion. I don't want to give away their system, so I'll tell you how I apply this idea more generally, in my regular physics class.

I often give students a 0, 1, or 2 simply for having sufficiently addressed all sections of a lab write up, from *Problem* through *Conclusion*. I will have anywhere from three to five such criteria. Another criterion might score 0-2 points for correctly processing the data, including a graph when relevant and a numerical result. I often will assign a score to the conclusion as well.

Next, I assign a percentage score for all possible combinations of numbers. For example, if you have four criteria, the max score is an eight. Obviously, an eight is a 100%. But should a 7 be an A or a B? It's up to you. I'd probably make that a 92%.

My rubrics then become flexible based on which criteria I grade. You can make up your own criteria. Think way outside the box. With NGSS in mind, you might have students address a cross-cutting standard in their conclusion, say *patterns*, or *scale and proportion*, whatever is relevant. How they address the cross-cutting standard can be assessed with a 0,1, or 2.

In essence such rubrics are numerical scales with graded criteria as placeholders. Picture two columns of numbers. The numbers on the left are the total possible based on the number of criteria you are addressing. The numbers on the right are the percent equivalences you assign to the score on the left. I go to 100%. I have several pairs of columns like this, one totaling to 6 points, and others for 8, 10, and 12 point totals.

Projects. Depending on how you set these up, projects can have even more moving parts than labs. But you can simplify grading these with rubrics too.

Here is one way to grade the Anti-Entropy device from Chapter 4. Again, here were the criteria:

5. Make the device.
6. Demonstrate it.
7. Make an annotated sketch of the device.
8. Include a paragraph explaining a) how the device decreases entropy when you work the system, and b) how the entropy in the universe *increases* even though you've *decreased* the entropy of your system.

Each of these requirements can be graded 0, 1, or 2 resulting in a total possible score of 8. You could divide 100 by 8, to let each point be worth 12.5%, but I think that's a bit harsh. Maybe each point is worth 8%. Or, you will consult your general rubrics, with point columns already matched with percent scores.

What made this particular project super easy to grade is that I could evaluate this as the students presented. Also, the grade was a group grade, unlike labs, which are individual.

Group grades are nice, because if you have a group of 3, that's one-third the grading required. If you think there are some leeches, probe them. Make other people answer the difficult questions. Scale the group grade down if they flub it. People will catch on that they all need to know their stuff. Also, don't give separate grades to separate group members. They stand or fall as a team.

Remember also, you might want to use the quiz option to scale individual member grades up and down.

Finally, there is something about "no evidence for," "partially complete," or "complete" that bugs me. What about excellence? As I said earlier, on this particular project I actually gauged audience visceral reaction and awarded +/- up to 4 percentage points for entertainment value, thusly determined.

TIME SPENT GRADING

Like I said before, you can punish yourself with heaps of gradables. Don't. Diminishing returns come quick in the grading world. If a kid gets an F on Tuesday, he's gonna get an F on Wednesday. He might think you are just rubbing it in.

Follow my guidelines above. There are exactly two things you need to keep in balance: giving feedback that informs, and preventing your life from falling apart. Value each, and you're a peach.

There are some things I cannot give you the key to. One of those is processing speed. Your brain will achieve maximum processing speed at the age of 25, but it will not see its peak in grading speed until you are in your 50s. Really, you need the experience.

So take heart, there will come a day when you will sit with stacks, and your left hand will be sorting as your right hand is marking. You will look like some insane red-ink Rain-Bird sprinkler, and before you know it, the stack will be gone. Then you can go home and watch re-runs of *Breaking Bad* with your husband, or do origami with your wife, or motocross with your high school buddies, or whatever you want.

Seriously, one of your professional goals has to be doing this stuff quickly. It very well may save your career. You don't have to wait until you are 50. Work on attacking the stacks. More in the burnout chapter.

Summary:

- **Don't be too harsh.**
- **Don't be too easy.**
- **Shoot for 15-25 graded assignments per semester, per class.**
- **Be as instant as possible in feedback with everything, tests and quizzes especially.**
- **Make and use rubrics to give good feedback and speed up your grading, labs and projects especially.**
- **Be quick, save yourself time.**

13. Peers

IT'S SO OBVIOUS

If it weren't for my buds, I would have been out of teaching the first year I started. And, by *buds*, I mean teachers I met on the job.

People need people. That's no secret. It's so obvious that I feel
foolish even having this chapter. But I know if there are other people out there like me, they need a how-to book that reminds them of this.

I would never have stepped into Helen Simmons' science room at Weller Middle if I hadn't been desperate for the help, and humble enough to ask. I'm sure I would have sidestepped a lot of trouble had I asked her help earlier.

In retrospect, I was the stupid robot. She was the connected empath, the happy Buddha—Ghandi meets P.T. Barnum meets daddy coming back from war—a respite from the horrors of life and the soul-shearing pain of middle school adolescence. I was so out of touch.

In the end it worked out well. And, while she influenced how I taught, it was more like giving this rock a nudge to get me rolling again. We met over several lunches. My biggest takeaways were what she thought about project based learning, and her easy, happy demeanor with kids. I am grateful to this day.

You might think you have plenty of friends from your teacher education program who could support you and give you ideas. Maybe, but having someone at the school has the advantages of being there when you need them. More importantly, they know the student population. They are in the trenches with you.

It's probably like coaching in professional football. Different teams have different attitudes, things they will respond to that other teams won't. Your on-site teacher friend will know these subtleties. The longer she's been there the better. Also, you want to find a teacher with a good reputation, not necessarily the principal's favorite, but maybe. The principal would be a good place to start asking, or the staff drunk. He's got nothing to prove. He'll tell you the truth.

LIFELONG FRIENDSHIPS

When I was at Heinlein High, I already had my sights on teaching in the IB program at Sakura High. But I knew my colleague from my credential program had it locked up (the 16 hour-a-day guy).

Then a summer school job opened up—just teaching geometry in a trailer. I couldn't make the first couple of days, so they put a new graduate on to cover. He didn't even have his credential yet.

I was grateful Bill covered, so when I came to relieve him I brought a fistful of cash, enough to cover the proportion of days he worked on my behalf. Maybe that made an impression, because we became best of friends after that. I was best man in his wedding just a few months ago.

Later, he introduced me to the works of Ben Graham. This new interest took hold, tugging me in the direction of financial advising. Years later, after I lost a series of administration positions, that's what I did. I came back to teaching a couple years after that. More on that later.

The next year I got a summer chemistry job at Sakura. Ken Petersen came in to my class as I was teaching and said he was moving north. He put my name in to replace him to teach IB Physics at Sakura. I was in. Wow, it just fell into my lap.

When I started that fall at Sakura, we had department meetings. They were real snoozers—the live announcements variety I mentioned above. But there were some electric young teachers at that school. At least a couple of them were new to the school too. We talked about getting things done at meetings.

So, as the meetings progressed, we introduced ideas that mattered, like improving student engagement, getting new equipment, and teaching better. Eventually, our 50% teacher attendance rate rose to 100% at these meetings. People wanted to be there.

We had a relaxed, working attitude. I made lifelong friends I met at those meetings: Savannah Rocco, Tabitha Breen and the Four Horsemen: "Famine," "Death," "Pestilence," and "Arsenal." (I was "War".)

We enjoyed our time so much, that once in a while, after school hours, we would go to a local pub and dream some more. We got a lot done.

Eventually, Dr. Rocco secured some grants allowing all the science teachers at the school to work with the feeder middle school teachers to create a coherent, strong science program in the community. We got even more done under her guidance.

But there were other things we did that served us well. Sakura was urban. We had a diverse mix of cultures and languages. We had a lot of Spanish speakers. We thought it would be nice if we science teachers got together socially once a month or so just to hang out at someone's house. We had only one rule: Speak Spanish. So, no matter how bad we were at Spanish, that's what got us through our "Noches de Espanol". They liked to drink wine at those meetings. You would think that helped to loosen the screws at the back of the tongue. But the truth is, it made me stupid. Still, it was worth it. We built relationships, and we got better.

What we did there lasted. I interrupted my connection with a ten-year odyssey, going into administration, leaving the district, then leaving the profession. Towards the end, Savannah called me back for a visit to help her move her stuff. She was on her way to Europe to teach in a military school.

I showed up in the afternoon. She had her cherished gear in plastic boxes, stacked up on the sunset-reflected wall of the now not-so-new science building. She said it felt like old times. I helped her carry her stuff to the car. She was on her way out. Unbeknownst to me, I was almost back in.

These relationships were nice, but they also helped professionally. We planned. Persevering, we brought about positive, empowering change. To me, the most appreciated thing was the fluidity with which we operated. We did favors for each other. And we made friends. Maybe that is the most important thing.

Summary:

- **Get to know your peers.**
- **Ask for help.**
- **Be positive in meetings.**
- **Look for people to dream and plan with.**
- **Be open to new relationships, and they will come, to your lifelong benefit.**

14. Lab Equipment and Supplies

Imagine you had no equipment whatsoever. Could you teach biology, chemistry, or physics? My friend Tabitha Breen said, "I honestly started with nothing and taught for weeks using my imagination—now I'm back to nothing again."

Some years ago in one corner of the country there was a movement among Advanced Placement Chemistry teachers to stop doing labs entirely. They felt there was so much they had to cover that labs got in the way.

I understand the need for job security, but that's ridiculous. Labs are worth it for labs' sake. We want to be doing science, not just telling about what other people did in history.

Besides, labs should improve learning. When I think back to high school, I remember the labs and demos more than just about anything else. I can't imagine teaching about acceleration without having students measure distances and times.

When the SciTech Academy was new, we had guest speakers from an Industry and Education group. They emphasized there were three things needed of tech workers—not programmers or front-end developers—but workers: the ability to show up on time, the ability to communicate well, and the ability to measure things well. Therefore, whether people go on to college or not, fiddling with meter sticks, stopwatches, and graduated cylinders is worth it.

This chapter is merely a short shopping list for things you need to buy, the common pieces of equipment and supplies you will need, once you start doing labs. The good news is you probably won't have to shop for these. Just look for them. They are probably sitting in a cupboard or drawer in your classroom or stockroom right now.

If they are not there, maybe someone pillaged your room when the former teacher left. Or maybe the previous teacher took them when he got a job at a different school. I once knew a guy who thought it wasn't stealing, as long as the stuff stayed in the district. I disagree, and most folks are honest, but cover all your bases. Make sure you have the basics. If you don't, talk to your peers and ask if you can borrow or share. Maybe there is a central place where this lab equipment is kept. Or maybe someone knows where *your* stuff is!

These are not exhaustive lists. They are the bare minimum. But if you had just this, you could get some things done.

Some districts have equipment that came with a textbook purchase. Sometimes there is a separate set of equipment that goes with a certain program for doing hands-on science. In these cases, there might be specialized pieces.

For example, when I taught in a different district, they introduced a piece of equipment that was an all-in-one physics of motion apparatus. They bought class-sets for each physics teacher. The pieces worked well. When I came back to my current district I saw one of those contraptions sitting alone in a corner. No one was using it.

Years before that, they built a new science building. They stocked it with so many hotplates that if we found a way to plug them all in at once we would have blacked out the city.

Here's the stuff:

PHYSICS

Equipment

- Glassware, various sizes, including beakers and graduated cylinders
- Stop watches/timers (Now we just use students' smart phone timers.)
- Meter sticks and rulers
- Spring Scales
- Light Meters
- Sound Meters
- Lasers, hand-held
- Hotplates
- Thermometers
- Multimeters & Wire Leads (voltmeter, ammeter, preferably including AC)

Supplies

- Batteries, 9V, 6V, 1.5V
- Light bulbs and sockets
- Safety Goggles
- Protractors
- Duct tape

These are short lists. To some teachers these are laughably short. But if I had these in place, I could overlay my patented "toothpicks and Styrofoam" strategy on top and teach anything.

You will also notice how many of these items are measurement instruments. Light meters and sound meters are likely not at most schools. And they are not cheap either. To me they are essential. They open up analysis to include more variables. If nothing else, think meters, meters, meters.

A simple light intensity lab examining I(ft-candles) vs. r(m) relates to many other physics concepts, from *Gravitation* to

Coulomb's Law to *Olber's paradox*. See the cross-cutting standards in Chapter 5.

You will also note the absence of the Monkey-and-Hunter demo, a go-cart, frictionless carts and tracks, and other things I've mentioned elsewhere in the book. You can get these over time, but start with the basics. I consider graph paper essential too, but you can photocopy from downloadable grids.

Other items will come to you. Instead of carts and tracks, you can ask your students to bring in battery-powered toys, which you can use to measure speed. If you ask for bowling balls, you will get these too. All you need is one though. Don't forget to bring in your tortoise.

CHEMISTRY

Equipment

- Glassware, various sizes, including beakers graduated cylinders, test-tubes, stirring rods Ring stands and test-tube clamps
- Hotplates and/or Bunsen-burners
- Thermometers
- Dropper trays and droppers
- Burettes
- Balances (scales)
- Scoopulas
- Ice-maker
- Gas cylinders, with various gases

Supplies

- Hydrochloric (muriatic) acid
- Sodium hydroxide solution or pellets
- Acid/Base ph test strips, litmus

- Distilled Water
- Salt(s)
- Magnesium Ribbon
- Safety Goggles

BIOLOGY

Equipment

- Glassware, various sizes, including beakers graduated cylinders, pipettes, stirring rods
- Petri dishes
- Microscopes
- Dissection pans
- Scalpels
- Balances (scales)
- Hotplates
- Thermometers
- Meter sticks and rulers
- Models (skeletons, etc.)

Supplies

- Microscope slides
- Iodine solution
- Glucose
- Dialysis tubing
- Acid/Base ph test strips, litmus
- Safety Goggles

These basics can cover a lot of labs. Conversely, a lot of labs require these basics. If you want some guidance about appropriate labs to run, I suggest putting all the words in each list in Google search and see what comes up. Don't forget to add the word "school". We live in a great age!

Nevertheless, I think it would be even better if you asked your experienced peers.

Summary:

- **Labs are important.**
- **When you get to your site, first dig out the items on these lists.**
- **Put this equipment and these supplies in prominent places.** (But not necessarily immediately accessible to students. Think safety.)
- **Search lab ideas from the Internet and from peers.** Don't neglect what your district suggests. Schedule the labs your equipment permits.
- **Be on the lookout for new labs you really like.**
- **Start diverting available funds to equipment and supplies you need** to run these new labs and maintain the core.

15. Field Trips

Most of the time, YOU can be better than any field trip you could ever find or that ever presents itself. You have so much available today: video, equipment, and persona. But field trips might still be worth it.

Here are three values that need to be paid homage to: Improving test scores, changing the world, captivating student interest. Here is why I value each:

REASONS FOR FIELD TRIPS

Improving test scores: This is the currency of the educational world. Yes, in many regions the move to national standards has suspended our use of tests, but this is America, and we like results. So, tests will return. I also believe that tests can reflect learning. I also have a need to prove myself, even at my age. I love breaking personal records.

Changing the world. Face it. There are reasons we do science. Hawking says we need to preserve the species by diversifying habitats across the universe. Kurzweil says we can live forever. Global warming/climate change/ocean rising IS REAL. How can we participate?

Captivating student interest. See immediately above. See the Monkey-and-Hunter-demo. See motivation research. See.

I WILL NOT WASTE MY TIME if a field trip does not offer one of these connections. I will take serious notice if it hits all three. I will be less inclined if it takes money away from equipment and supplies. Therefore, if you find a charged message by visiting a certain place, and if you get grants to pay for the bus and the entry without having to jump through hoops, you've got something.

Here are two good examples:

San Diego Gas and Electric Energy Innovation Center

San Diego Gas and Electric (SDG&E) has a conservation/renewable energy education program for all ages. The high school Energize Schools program is run by the non-profit organization Strategic Energy Innovations (SEI). They use the SDG&E Energy Innovation Center as a learning laboratory. It's a building full of relevant green building, energy efficiency and renewable energy design features. There are giant overhead solar panels providing shade to cars in the lot. They track the sun as it moves, yielding maximum power. The panels look like big flat trees.

SDG&E cools the facility by freezing a few-ton block of ice at night, when electricity rates are low. As the block melts throughout the workday, cold water circulates through facility walls. Big slow fans nudge cool air from the walls into the room. A garden features a wall of plants. This too is a design feature. It insulates and beautifies buildings.

SDG&E provides buses for tour facilities. They also have an end-of-the-school-year fair, where students can exhibit energy projects. There are also adult speakers. The day includes a revelation of the school winning the energy conservation prize and cash award.

If a school engages in work with SDG&E, they provide teachers with quality curriculum teaching electricity concepts. When you add it all up, the SDG&E field trip scores a three out of three: Test score help, world changing message, and meeting student interest. They rack up extra credit with their free buses.

South Park Renewable Energy Center

Part of the IB curriculum focuses on energy and renewable sources. Every year I take my students to a local house that has a 220V electric car charger and a pool heated by a solar panel. We start by gawking at everything.

We bring thermometers, giant graduated cylinders, and stopwatches. Students use these, going to work measuring water flow rates entering from the passive solar panel on the roof of the garage. They measure temperature rises over time. They compare this to the known rate of power from the sun. In good years, they think to adjust for the atmosphere and the sun's angle.

Some of them swim. We all get burritos from a local taco shop. Last year we made vessels to transport the owner's tortoise across the pool. Parent volunteers have as much fun as the students.

The experience becomes a lab write-up satisfying IB requirements. We also have a few of the students prepare a report for the SDG&E exposition.

Let's do an accounting:

Test score help, CHECK!
World changing message, CHECK!
Meeting student interest, CHECK!
Low cost—we walk there, CHECK!

You will need to consult your principal's secretary to make sure you dot the I's and cross the t's. There are two forms you'd better get done, or it ain't happening. First, you need to get a form that requests all teachers' signatures for the periods your students will miss on the day of the outing. Second, you will need to get parent permission for the trip.

That's not TOO burdensome, but you should make sure you get all signatures AT LEAST two weeks ahead of time. It gets more complicated when you are dealing with seniors and kids over 18 who drive, and other kids who want a ride with them. Talk to the admin assistant about this mess.

Also, you will need to check with your district policies on how many parent (or other) volunteers you need to get to chaperone a field trip. It will be one adult per 10 to 20 students.

One of my favorite side effects of field trips is being able to talk with other adults. Often, they are stay-at-home moms or retired people or independently wealthy folks who can do what they want when they want. Sometimes you get working parents who take a day off. Whatever the situation, take advantage of the opportunity to have stimulating conversation on the bus.

AWARENESS

I had a great field trip my first year at Weller Middle. I was taking 30 kids to something, but 40 got on the bus. Those little rascals helped themselves, and I didn't expect it! I didn't make a very good impression on my admin. Maybe that's why my VP gave me a substandard evaluation.

Make a list and take roll before you let them on the bus. If you see stragglers, look them straight in the eye, and say, "Get back to class!"

Summary:

- Take students on field trips if the trip does most of the following:

    1. Helps with test scores,
    2. Works for a better world,
    3. Piques student interest.

- Make special consideration if you can easily get to the destination without cost.

16. Extra Curricular/ Volunteer

"We're getting an energetic young science teacher. He's going to do that Science Fair program we've wanted to get going. Parents are going to be so happy!"

That's the principal and the vice principals talking about you. I'm pretty sure this conversation happened about me the day I got hired at Weller Middle. I'm glad I didn't know about it at the time.

The principal was nice. When she finally asked me to do science fair, what could I say? But I chose instead to focus on the compound-emergency that was unfolding on me that first year. By "chose" I mean I let the urgent needs take me over. Procrastination won. It had to. The urgent need was too important.

I made the right choice. I tamed the class beast. And I even got some of my better students to slap together a couple of tri-folds for the pre-judging. The principal let me know she wasn't happy. It was weak, but at least it was a showing.

The next year, I did it right. We had a lot of projects and some good qualifiers. She was happy then.

So, if you get asked to lead one of the following extra-curricular events, I recommend you do it better than I did, but I also recommend you also prioritize as I did: class control before anything else. That's the most important thing.

The way you get better than me at all the rest is simply to ask. Ask the other science teachers on-site what the history is for that event on campus. Email or call the local coordinator for Science Fair or Science Olympiad. They could help you immensely.

If you don't get asked to do any of these, count yourself lucky. But by year three or four, you might want to take one of these on, at least as an assistant coach.

SCIENCE OLYMPIAD

There are a lot of science-specific extra curricular opportunities for students. I like most of the ones I've seen. Science Olympiad has student teams compete on Saturdays making devices or taking tests. One event has students build an analog device that accurately takes time. Students make sand clocks and water devices, pendulums, whatever.

Other events include bottle rocket launches. A set amount of air pressure is pumped into a 2-Liter bottle partially filled with water. Winners have the longest hang time. Hang time becomes a function of height achieved, and parachute effectiveness. My colleague, "Arsenal" once took a middle school team to place in the national competition.

It is a great opportunity to work with students completely focused and bought-in to doing something special. Talk about motivation. Just get out of their way!

ACADEMIC LEAGUE

Panels of students answer questions competitively, smacking timers when they think they have the right answer. Questions could be in any academic field, including current events. If someone on the team answers correctly, the team gets a chance at more points in the bonus round. There, the team gets a short time to answer a series of two to five questions.

SCIENCE FAIR

Students do science projects that get judged at the school. If they make it to the county fair, they get a shot at going to the state competition, which could have cash awards and application fodder for college. There are decent cash awards even at the county level. This might have played a role in my own son getting into DESMA at UCLA.

WHY YOU WOULD WANT TO PARTICIPATE

The first time I did a competition like this, I found myself getting swept up in the excitement. My team was doing well on that Saturday. I was getting real aggressive advocating for my kids. I had to check myself because I was going over the top. It was fun.

Sometimes the school offers stipends for coaching some of these events, but not always. I wouldn't do it for the money anyways. It is just fun, and it builds interest in science. It also builds connections with students and parents.

Other times the school values the program so much they schedule you with a teaching period dedicated to that competition. I have never had the chance to do one of those, but that would be a great opportunity.

ASB AND HOMEGROWN CLUBS

Sometimes you have students with unique ideas for clubs. Or you might think of something interesting to do on your own. I was talking with a law teacher at our school last fall. Percival had a great alternative to the boring old stock club. He thought it would be stimulating for our students to go one step further and have our people speculate on *derivatives* of stocks, specifically *puts* and *calls*. From that, the derivatives club was born, co-sponsored by Percival Green and me.

They were not going to risk any real money, but you never know! If they proved to be stock geniuses, maybe we could leverage the school foundation money up to stratospheric heights, by merely advising the foundation.

We came up with a mission statement:

> Derivatives club: to gain facility with the employment of financial weapons of mass destruction for the common good of those who would wield such instruments and their future bosses, those who cross oceans in robotic yachts or associate with gangsta rappers who wear so much gold they could sink such yachts.

We learned about puts and calls at lunch on Mondays. Then, students set up sample portfolios of their picks on a finance site. Each person defended his picks. Each week we saw the results and teased the failures and lauded the champs. Some students had phenomenal results.

Just last week we found out one of our guys got second in the county in a yearlong stock-picking contest. This was a surprise to us, because after each of the twenty students set up their portfolios, we all forgot about it.

We turned our attention to the derivatives. Then out of nowhere we got representation at the culminating stock event. Hopefully our winner will get a chance to talk about the yachts, the gold, and the gangsta rappers.

Summary:

- **Extra curricular activities are good for the school reputation.**
- **They attract families and motivated students to your school.**

- **You might be asked to lead something when you are new.**
- **If you are, ask for help from peers or the local organizers of your activity.**
- **If you are not, wait until your third or fourth year teaching,** and then consider participating, either as a founder or an assistant coach.
- **Clubs can be fun,** especially if you want to start something new.
- **Students, administrators, and parents will appreciate anything extra that you do.**

17. Parents

In the fall of 2011 I was two years out of education and two years into my new career, financial advising. I had abandoned education after an eighteen-year career the second time I lost an admin job.

I was doing well. I came out of the gate "exceeding" as my brokerage firm labeled it. But in the summer of 2011, S&P downgraded the U.S. long term rating to an AA+. At that point my business froze solid. No one was investing for months.

That's when I got a phone call from my pal Tabitha Breen. Sakura High's Walk on Water IB Physics teacher left for a job up north. So from Labor Day to mid October, the students had a substitute. Granted, it was the very capable Percival Green, but he was a law guy, not a physics guy. They wanted me back.

The call couldn't have come at a more leveraging time. I was about to get on the bad list for meager sales from my brokerage, and this position was open at the school. I knew that every day was valuable towards preparing students for their May IB Exams. Half the fun of teaching IB was to shoot for high scores. So I had two good reasons to commit to one path and reject the other.

It made me mad. I was truly angry. Not at Tabitha, but at the situation. I did not want to leave what I had built in such a short time, but it seemed the logical choice. I went back to Sakura. I left my office on Friday. I was back in school on Monday.

It felt right. I got to know the kids quickly. They were bummed that they were without a physics teacher for six weeks, but I assured them we would get up to speed.

Something strange had happened in the two years I was out. I came back to realize I liked the kids even more than the curriculum. It was as if my two-year hiatus was a long-needed sabbatical. I also thought that maybe the high degree of face-to-face interaction and frankly, the sales experience I gained helped me connect with the students.

But the pressing need was scheduling the labs and activities that would prepare my seniors for their spring exams. The next most important thing was scheduling my juniors to get them on pace to step smoothly into their second year of IB Physics. The third priority on my list was to get the freshmen physics kids engaged and motivated to do science. I wanted them to consider IB Physics as an option.

So, I was spinning a lot of plates the minute I stepped back in. Then there was *AVID, Advancement Via Individual Determination*. Avid is a program that provides supports to students who are normally underrepresented on the march to college. It is an effective course that teaches students note-taking skills and other habits so they can do well in school, college, and career. It's a great program. I taught the freshman version of AVID.
As I saw it, the administration needed to fill my schedule with one more class, since there weren't enough physics courses to keep me going. I respected AVID, but it was last on my list of priorities. I felt I could serve the kids fine for a few weeks at that level of focus, then ramp up later.

But a particular parent didn't think so. She wondered whether I knew all the AVID strategies, because she wanted to give her child the best shot at college. I get it. When we had our first child in high school, Grade Nine, we were probably even more probing into the function of the school. But I was in no mood to change focus from physics to note taking skills. It was irritating.

Things eventually worked out. AVID turned out to be fun. We discussed the *Antifragile* idea from Nassim Taleb's book of the same name. The concept explains how some systems and people can not only be immune to hardship, but become better despite it, kind of like a human version of my anti-entropy device. Students came up with all sorts of examples. I remember one person bringing up Batman as an example of someone who is antifragile, brilliant.

But that is about all the parent friction I've had in my last ten years of teaching. And even that worked out. This parent has been gracious ever since. I had her child in physics later, and things were good all around.

I count myself very lucky, because if parents do not like a teacher, things can get difficult for that teacher.

PARENTS ON YOUR SIDE

Parents talk. They want their child in the best classroom to maximize their potential for success. As soon as they see that is your goal too, and as soon as they perceive you as someone who can deliver, they will be happy with you. They will talk about you positively.

In the opposite extreme, you could get labeled a lunatic. That's not good. I know all this by being a teacher and an administrator. I also know it because I was a parent of three, now college age.

I don't like being observed. I could be Alex Rodriguez stepping up to the plate opposite a tired pitcher and I still wouldn't want people looking at me. Therefore, to be my best, I try not to think of parents when I teach. The experience is for the students and for their understanding. If I imagined a mom or dad standing behind each kid, I would become extremely boring. It might benefit you to visualize it, but not me.

However, there are plenty of times during the school year when you can connect with moms and dads: open houses, Academic League matches, and when they volunteer to help proctor tests. You can tell where you stand by your interactions with them.
You can even form friendships. One of my parents is now my chiropractor. I have started going to concerts with another.

If one of your early goals is getting parents on your side, just do the right thing. This book outlines the right thing. Think of any chapter. Take Chapter 3: Fun. If students are having fun, imagine the dinner conversation. Take Chapter 4: Motivation. If they are motivated, they are going to do their homework. Your book open on their desk = victory. Take Chapter 5: Educational Theory. If the kids tell mom their brain hurts, but they learned so much, what mom wouldn't want a funnel full of knowledge going into their kid's head?

PARENTS NOT ON YOUR SIDE

This usually happens for one reason: fairness, or the lack of it. Did the student have a B two days before the grading period ended, but she got a D on the report card? Did the kid do the extra credit, but you lost the paper? Did Sally get to do a make up lab on Thursday, but you denied Rosa on Friday? Apologize, fix it, and move on.

Think of your job as a sales job. You sell to students, but parents call on the customer service line. Your company will be more profitable if you can dedicate more of your employees to delivering on sales than to putting out fires.

One last word, it is true that you cannot please all the people all the time. But if you have 95% of the parents on your side, and half of those are rabid advocates, who is going to look like the lunatic? Be part of the 1% of teachers, and you will have no trouble.

Summary:

- **Do your job with conviction.**
- **Be fair.**
- **Apologize, fix, and move on.**
- **Be part of the 1% of teachers, and you will have no problem.**

Chapter 18: Dress

I don't give a rat's ass about what I wear. Very recently, I went through a period of time where I couldn't help but imagine my face was made of angular black metal plate. I don't think I'm trying to hide. I just like efficiency!

I like how my son dresses. He has shiny jeans rolled up at the cuff revealing shiny black boots. He has a button-down solid bright-colored shirt or a T-shirt covered by a light sweater. He is much more tasteful than I am. I could teach in that. But I don't want to be bothered.

Here is a picture of three of my current students, all eighteen, so yay! I can use the pic. I like how Jarod is dressed. I love this pic. It could be an album cover. I should dress that way, but I don't.

THE POWER OF DRESS

Back when I was a Science Administrator, our biggest trick was to take all the time we needed to make awesome lessons, then deliver them. Being administrators, we wore ties (well, the woman administrators didn't necessarily wear ties, but they dressed the equivalent.) In other words, we dressed like teachers do in Texas.

One of my science administrator compatriots tells the story of the time he went into another teacher's classroom. The class was always way out of control. Not much got done.

Later, this administrator delivered a spec-ed out lesson to this same class, and things went very well! Students learned.

Afterward, the regular teacher said, “Sure kids listened to you, because you were wearing a tie!”

Dress matters.

WHY I DRESS LIKE THIS

Now that I’m back in the classroom, I say good-bye to the tie. There is one social studies teacher on campus who wears a full suit every day. He is a friend. He is also Canadian. I don’t know if there is a connection there, but none of the Californians dress that way.

Hate is a harsh word, but I hate stuff that squeezes my neck. No top buttons. No ties. I can breathe. I am active, so I don’t want a jacket, and I don’t want a white science lab coat. Nor do I want tie-dye lab coat.

Here in So-Cal, the other end of our dress spectrum is jeans and T-shirts. Nope, don’t want that either.

I dress for three reasons: 1) comfort, 2) action, and 3) not standing out. I am a science teacher. I am already weird. So I aim for the middle: Dockers pants and polo shirts, or golf shirts, maybe a button-down shirt. I either wear dress shoes or minimal-sole versions of the same. No tennis shoes or sandals. I’m not Jesus.

Let me make a distinction about not standing out. What I’m saying is I think there is value in looking like the typical teacher. If you are under thirty, you can easily be identified as a high school student. I’ve seen it many times. And if adults see it, students might see it that way too. They are better at telling age, but if your dress matches theirs, you will lose a little respect. You need all the respect you can get when you are new, so strive to look like the average teacher, not the average young person.

Maybe a dozen times in my teaching career I've been told, "You dress so well." Probably it's because I go that extra step to iron my clothes if things are wrinkly. But 12 times in 21 years is about right. I'm in the middle.

A couple of months ago I was passing out papers when Jose (the roach kid) told me my shirt was inside out. Well, what do you know, it was! And it was one of those polo shirts, so the collar was messed up too. I had rushed out of the house to get there in time, and that was a result. You can see the measure of care.

Ladies, I cannot advise you, but I assume you will be able to extrapolate from the description of how I dress.

Summary:

- **Dress for comfort.**
- **Dress for action.**
- **Dress to not stand out.**
- **Dress as an average teacher.**

19. Halloween

Question: What day has the highest percentage of strangers having sex? You guessed it. People go crazy on Halloween. There is a lot of excitement around this one day. So, as a teacher, do you want to dress up? Unless you have a strong religious conviction against it, I'd say YES.

Chapter Three was about fun. Are you really about fun? Then show it. Yes, you might lose some instructional time, but it is worth it.

You know all about my first year. I was struggling. Classes were out of control. By October 31 I was aware of this. I can't remember, but maybe I was trying to fix it by then. Regardless, I made the incredibly rash decision to let it all go.

My wife thought it would be great if I dressed like the Genie from Aladdin. We bought a whole lot of blue body paint and glitter. She sewed me puffy blue genie pants. I cut the sleeves off a tank top. That's how I showed up to Weller Middle School.

Guess what? My adopted advisor Helen Simmons showed up as a giant Hershey's kiss. Her hips were fifty-inch wide aluminum and the rest of her visible body was painted dark brown. We were a pair.

Had I not dressed, I would have been three times the doofus I presented the day after Labor Day.

So what did I do the next year? I stuffed some gray sweats with towels and came as Hans from the SNL pair Hans and Franz. I taught the whole day in a German accent.

I can't remember if I missed a Halloween since.

My second year at Heinlein High I was super-busy. We were preparing to go on strike and my wife was having our third baby. Plus, I was still new to teaching, at least at the high school level. I was thinking of passing on Halloween. But the morning of, I decided to be honest and go as the tired dad.

Paige put dark eye shadow around my sockets. I put on pajamas, a robe and slippers. I got a very realistic baby doll and put her in a car seat. I went to school that way. At the start of each class I told the students I was waiting for my wife to pick up the baby, and that I needed someone to watch my child. All the girls volunteered, hands up, so enthusiastic.

I chose one and brought her my baby. I tripped. I launched the child across the classroom. They screeched so loud! Those girls were so angry! I made them swear they wouldn't tell the next class. I pulled it off five times that day.

A couple of years ago I came as Walter White from *Breaking Bad*. Last year the show ended, so I wanted to pay my respects one more time. But I couldn't do the same costume. Instead, I was going to reproduce the first episode version, where Walter was wearing nothing but an apron and underwear. Hey! I was going to wear athletic shorts to take the edge off, but my wife wouldn't let me. She forbade me to do this costume.

So I said, "Heck, I'll just dress Griffin up like Tortuga from a later episode." And that's what I did. So my pet tortoise was my Halloween Avatar, recalling the episode where the California Desert Tortoise was carrying the head of a drug lord—and a bomb.

Students loved Griffin anyway, and they got to see him again. Each of the days I dressed up, I ended up teaching something, even if it was in a German accent.

Summary:

- **Yes, fun is more important than educational theory.**
- **Be fun. Dress up on Halloween.**

20. The Copy Room

Yesterday morning I hiked up to the staff lounge to make copies for my next class. I opened the door to find my one hundred pound, fifty-year-old chemistry teacher friend Shelley hip deep in a machine, flat on her back.

I thanked her for being so helpful. I congratulated her little hands, which were perfectly designed to pull shards of paper from intricate machinery. She said she needed forceps. I asked if she were delivering a baby. Then she asked for tweezers.

It's just a day in the life of a teacher. I lied, or rather, I forgot to mention in chapter 14 that the most important piece of lab equipment is the copier.

Remember the graphs I thought were so important for my eighth grade students? I needed to make some graph paper that morning. Turns out graphs are indeed important to science. They show the relationship between variables. If that isn't Western-Sciencey enough for you, I don't know what is.

There are tons of other documents you need in addition to graph paper. At Gigantic High they made so many graphic organizers that the oceans are at least a millimeter higher today.

You will learn to massage your copier. The first time you experience a jam, don't panic, just follow the prompts on the LCD display. You will open the front panel and see some hieroglyphics, which will further lighten your path. Eventually, you will clear your first jam and you will feel very good about yourself. And you will have salvaged your lesson.

YOUR ADVANTAGE OVER VETERAN TEACHERS

If you get there at 6AM and stay until 5PM, you will have no problem at all. In the worst-case scenario, if the copier breaks, they will call for a repairman. The repairman comes as part of the rental/purchase agreement for the machine, so he is usually there the day of the tragedy. So, if the copier is dead in the morning, it's resurrected by the evening. If it's dead in the evening, it's back in the morning. You win.

Of course it's best to plan a couple of days ahead, but I don't. One day is enough for me. But I am no longer a 6AM to 5PM guy. One important safeguard I have learned is to get to the copy room early in your prep period.

You can do this by paying attention to two things: 1) your yearlong plan, 2) what you did last year at this time. Think about these two things at the start of your prep period, and you will know what you need to copy for the next day, even before you put detail in your lesson(s). Knowing that, you can take your copy items to the machine early in the period, and you will beat those who wait until the end of the prep period, when they are all planned out. That's the majority of teachers. And, since you are a science teacher and understand how things work, you are not likely to break the machine. All will be well. Furthermore, if you are a new teacher, you have the early morning and late afternoon to get your job done.

Summary:

- **You will need your copier.**
- **Do not shy away from fixing it.**
- **Be the first in on your prep period.**
- **If you are new, you might have to work long hours, but you and the copy machine would have the early morning and late afternoon all to yourselves.**

21. Burn Out

In my eighth year teaching IB Physics at Sakura High, I found myself standing in the middle of the room. I said to myself out loud, "I'm done."

Even though I loved the curriculum and the clientele, I was bored. I was a little tired. I had been asked a year earlier to be science administrator. I turned it down. Now I thought I could take what I learned and teach others. So, I accepted the position. This little change of pace may have kept my education career alive.

And yet, had I stayed in the classroom, I would have kept doing the same old thing and gotten really bored. Kids don't deserve a bored teacher.

I didn't have to become an administrator to combat boredom. In the spirit of James Altucher's ten ideas a day, here are ten other things I could have done:

1. Connect every physics lesson to biology.
2. Connect every physics lesson to chemistry.
3. Make 50% of the labs student designed.
4. Go full Project Based Learning.
5. End each unit in a poetry competition.
6. Center the curriculum around world enhancing themes—a new theme each unit.
7. Center the curriculum on clarification of ignorance.
8. Connect every topic to sports.
9. Connect every topic to space travel and the propagation of humans throughout the universe.
10. Rank the kids' top six hobbies and dedicate one per unit.

And bonus ideas!

11. Go top-heavy on higher level questioning—lots of evaluation and synthesis questions.

12. Fashion a year's worth of curriculum on Marzano's top strategies.
13. Do three big school-changing projects.
14. Get some hobbies or side projects.

By the time you are ready to burn out, your reputation should free you to do anything you want. Do something crazy but good—like these things—and you can reinvigorate your mission.

Also, if you've been plugging away with long days for too many years, find a way to back off and keep the quality. You've seen my new schedule. I have several hobbies that nourish other sides of myself and that probably spill over into my teaching: music, writing, and financial advising. I also lead a men's group every other week at my house. This diverse set of things keeps me balanced, sane, and engaged.

But these things are more than clinical therapy. They fill my life with unanticipated possibility, struggle, triumph, surprise, and spiritual connection.

Summary:

- **Overwork and boredom can burn you out.**
- **Burn out is a sign that you have to make a change.**
- **Anticipate and ward off burn out by taking on new challenges.**
- **Simplify your work. Get more efficient.**
- **As you gain mastery over your teaching, broaden your interests outside of work.**

22. The End/ The Beginning

VETERAN

There is a local administrator in town. He oversees a chain of charter schools. Every year he gives the same speech: "A fifteen year veteran teacher is identical to a first year teacher doing the same lessons fifteen years in a row."

That's an insult. And it's wrong. My learning spiked in year 1, year 2, year 3, year 5. I keep getting better. And I'm in year 21. There is research on teacher efficacy vs. time in the field. Some research points to a peak at year 23, with a decline after that.

Oh yeah, I could see a teacher letting it all go in his last year. I could make the whole year a party. But I want to go out on top. Maybe my 30th year will be a party, *and* my most effective year to date. That would be a good rhyme.

When I think of Tabitha Breen, Dr. Savannah Rocco, Arsenal, Death, and Stella Groves, I thank God for wrinkles, and the wisdom that comes with them.

There was a movie from the Seventies called *Logan's Run*. It was about an ideal futuristic society where everyone had a recycling ceremony at the age of thirty. On their birthday they got blown up and replanted into an infant. Friends came to clap and cheer their buddies as they got blown to smithereens. This was to keep everything optimal, maximized.

I won't spoil the ending, but they were wrong to blow up the 30-year-olds. If we had the same ceremony today, I think our society would slip back to caveman mode.

You are awesome. You are valuable. We need you. But an experienced you is better, because you keep getting better. Don't let anyone tell you otherwise.

PHASE II

When I was a financial advisor one of my favorite encounters was with a beautiful woman of fifty-seven years of age. She welcomed me, but didn't often have time to talk, because she was working hard on her new career, making maximally effective web pages for companies. It was a second career, and something she found she had a talent for.

She told me about a movement. It was called something like, "The Next Fifty". The idea is that you are not done at fifty, but you start something brand new. I like that. That's probably way off for you, but don't forget it.

ENDING IN STYLE

When you start as a year-one teacher, you don't make much money, but you have opportunities to put that money to work. And you should. Teachers are permitted to put away up to $18,000 a year in a 403B account.

Now, when I was a new teacher, I was also a new husband and a new dad. There is no way I could have put that amount away, and I didn't. A rule of thumb says that 10% of your income should go into retirement savings. But please, not *savings*, retirement *investment*. Ten percent of $40,000 is $4000 per year. That's $330 to $400 a month. Of course you could use that now, but I suggest you invest it. And, when you make more, increase proportionally.

If you are going into a 403B account, you will be lowering your current tax burden, so you will actually see less than that amount coming out of your check.

An additional option open to anyone making less than $100-and-something-thousand dollars a year is a ROTH account. I love ROTHs. You take up to $5500 a year from your bank account and invest it someplace, maybe with a discount broker like Fidelity or Schwab or Vanguard. You can buy stocks or funds. Since the money comes from pay that has already been taxed, the awesome feature of Roths is that the money will never ever be taxed again, even after it grows—even if it grows like crazy!

During the last few hundred years of the stock market, gains have been real and substantial. Yes, there were times when everything went to horrible hell. But the real and substantial gains factor that in.

Some people believe that this generation of new professionals will miss the benefits of investment because they were trained by a generationally bad horrible-hell stock experience early in their conscious life. They fear the market. Don't let this happen to you. With risk comes reward. If you want reward, you take risks.

Some people say, "You know, I think I could put that 10% to better use saving for a house, getting a car, and rewarding myself with a vacation." Trust me, all those things will come in time, but what people fail to understand again and again is that most of the power of investing is the *time* ***young*** people have for compounding—if they take advantage of it.

Here is an exercise you need to do. In your mind's eye, picture your favorite male or female symbol of beauty. See him? See her? Go on line and confirm that what you are recalling is true. Aren't they just awesome? Now think what they will look like at the age of seventy. Better yet, look up some stars who are now seventy. Take it in. Who would you rather hold hands with and walk down the street, the young or the old?

No contest. That same advantage of youth applies to investing. Compounding depends on time. You have time. Don't wait to start investing until your compounding potential has the appeal of a seventy-year-old.

Now, you might think that you should put off investing until you are established enough to afford it. You have a home, cars, stuff, so finally you can trickle funds into your account. Good for you! Working hard, you might have a hundred or two hundred thousand dollars in your accounts by the time you retire. Therefore, if you take out the commonly advised 4% per year, you can supplement your retired income by $300 to $700 a month. Can you go to France on that?

However, if you are like other diligent investors who start young and put aside significant sums, you and your spouse can retire with millions. You might have a $10,000 monthly income from investments alone! Compare that with $500 per month. All it takes is the diligence to commit to 10% from the time you are young. No guarantees on the $10k, but it could even be more. Will your income rise more than expected? Will your spouse have a high-paying job? The ten percent rule would smile upon these scenarios.

I can't tell you what to invest in. And I can't tell you how I've done. My status as a financial adviser limits what I can say. However, the power of compounding is real. Youth is a real commodity for compounding. Take advantage of it. Talk to friends and family and professionals—whoever knows something about investing—and heed good advice. Read a book! It's not rocket science.

TIME TO MOVE ON, AND UP

The happiest people walking the planet are retired teachers. I know, that's way off for you, and you are not even sure you are going to commit to thirty or more years of teaching. But let me lay out the advantages.

Teachers get summers off. Yes, a lot of them work other jobs, but the rhythm is different. Last summer I did my own thing. I went bodysurfing. I wrote my weightlifting book. I did nothing. It was great.

The fact that teachers have summers off when they are young gives them practice at being retired, and that's a good thing. You get in the habit of pursuing passions. That habit follows you into your retirement years.

I dedicated this book to retired teachers. I meet with one of them on a regular basis. We have great talks, reminiscing and dreaming. She still cares deeply about the school she graduated from.

Sometimes, when we are walking across campus, or from the car to the restaurant, I hit my usual robotic stride. Then I catch myself—I am walking with an 82 year old. But she's right there with me, not missing a step.

I believe retired teachers are not tired. I believe they are energized, because they've wrestled with cosmic forces their whole lives. And they won't let them go.

Made in the USA
Monee, IL
04 November 2026

67375675R00095